The Asbestos Racket

The Asbestos Racket
An Environmental Parable

by
Michael J. Bennett

with a foreword by
Reed Irvine
Chairman, Accuracy In Media

FREE ENTERPRISE PRESS
BELLEVUE, WASHINGTON

The Asbestos Racket

FIRST EDITION
Published by the Free Enterprise Press

Typeset in ITC Clearface on DTK computers and the National TeleVAR Turbosetter 1000 in the Don Hummel Memorial Workshop by the Free Enterprise Press, Bellevue, Washington. Cover design by Northwoods Studio.

The Free Enterprise Press is a division of the Center for the Defense of Free Enterprise, 12500 N.E. Tenth Place, Bellevue, Washington 98005. Telephone 206-455-5038. The opinions expressed in this book are not necessarily those of the Center, its board of directors, or its members.

This book distributed by Merril Press, P.O. Box 1682, Bellevue, Washington 98009. Additional copies of this book may be ordered from Merril Press at $9.95 each.

LIBRARY OF CONGRESS CATALOGING-IN-PUBLICATION DATA

Bennett, Michael, 1936-
    The Asbestos Racket / by Michael Bennett. -1st ed.
        p.   cm.
    Includes index.
    ISBN 0-939571-11-0 : $9.95
        1. Asbestos–Health aspects. 2. Asbestos abatement industry–United States. 3. Asbestos–Government policy–United States.
I. Title.
RA1231.A8B46   1991
363.17'91–dc20   91-30408
                                              CIP

PRINTED IN THE UNITED STATES OF AMERICA

*There is no naturally criminal class
in America—save Congress.*
–Mark Twain

Contents

The Asbestos Racket

DEDICATION
For Mike, My Father

Who will guard the guards?
                                                    –Tacitus

Author's Note

My father died of cancer. He had been a smoker, but had quit, but then he had worked with toxic chemicals for 20 years. At 68 he knew he would die sooner or later. But he wanted to know the how of carcinogenesis–if not the why of death: The latter he left to God.

This book is a modest attempt to provide explanations, if not answers, for all cancer victims and their families.

Two immutable realities–as stark as death itself–emerged from almost 20 years of study, research, and writing:
"Every food and every drink, if taken beyond its dose, is poison."
                    –Paraclesus (1493-1541), founder of modern medicine

"This stinking tobacco smoke. . .hateful to the nose, harmful to the brain, dangerous to the lungs."

                                            –King James I

And politicians of every stripe have never understood the basic concept:
"Above all, do no harm."

                                            –Hippocrates

For the politicians do what they have always done–subsidize rackets.

The Asbestos Racket

Like all books, *The Asbestos Racket* is the work of many hands, even though only one name appears on the title page. Dr. Aaron Wildavsky, America's foremost authority on risk assessment, deserves great credit for standing up to asbestos hysteria and speaking out against it; I found him an inspiration always, and a supportive friend during times when this project seemed like it would never come to fruition. The editors of *Science* kindly gave permission to quote from specific articles published in their journal. Others too numerous to mention provided leads, tips, interviews, editorial comments, and reviewed chapter drafts in progress; you know who you are and I thank you.

I would like to acknowledge Alan M. Gottlieb, publisher, of the Free Enterprise Press for accepting my manuscript for publication; my editor at the Press, Ron Arnold, for his unflagging encouragement and insistence upon thoroughness through many drafts; to Susan Perlbachs, assistant editor, for her yeoman labor editing, proofreading, and for constructing the extensive index. My deepest thanks to Reed Irvine, leader of Accuracy In Media, for his enduring interest in the asbestos controversy and his willingness to provide the introduction for this book.

Whatever merits this book may have must be attributed to these fine people. Any errors of fact or judgment are mine alone.

-Michael J. Bennett
Washington, D. C.

Reed Irvine
Chairman, Accuracy In Media

Foreword

The *Asbestos Racket* is a book written by a reporter—and a radical.

That may seem like a contradiction in terms, and hardly something I would endorse as the chairman of Accuracy in Media.

But it's not.

Mike Bennett is one of what unfortunately seems to be a vanishing breed of journalists, the kind of reporter that had tough city editors pound into their brains—and kick into their butts—one fundamental axiom:

"Never assume anything."

"Radical" means in *Webster's Third International Dictionary* definition "going to the root or origin" of anything, "touching what is fundamental."

And that's precisely what Bennett has done in *The Asbestos Racket*.

What he's done is force the Environmental Protection Agency (EPA) reluctantly and under great pressure from the scientific community to admit its asbestos removal policies were wrong.

And I think he's demonstrated, in the process, that the only good reporting is fair reporting, that assumes that there are at least two, if not three, four, or five sides to a story. And that makes much more interesting reading than the polemics that pass for reporting, all too often, on the pages and screens of our mass media.

Bennett has found in federal policy on asbestos a modern example of mass hysteria, as irrational as any of those described in the classic study, *Extraordinary Popular Delusions and the Madness of Crowds* by Charles MacKay, first published in 1841. MacKay's book examined the witch trials of the Middle Ages, the great financial speculations of the 18th century (including the Mississippi Scheme and the South Sea Bubble, which first enriched and then impoverished hundreds of thousands), alchemists, fortune tellers, and astrologers.

Follies and fads have dictated markets, medicine, governments and life throughout the centuries and continue to do so today. "Grand scale madness, major schemes and bamboozlement has been transforming sen-

The Asbestos Racket

sible, intelligent people into mayhem-making mobs" for centuries, as the dust jacket blurb on the most recent edition of the book, published in 1980, noted: "Remember Beatlemania."

Indeed, Bernard Baruch, the famous financier, writing a preface for the edition of the book published soon after the Great Crash of 1929, claimed it saved him millions of dollars, by illustrating—and demonstrating—one simple principle:

> "I have always thought," Baruch reflected, "that, if even in the presence of dizzily spiralling stock prices, we had all continuously repeated, '*two and two still make four,*' much of the evil might have been averted. Similarly, even in the general moment of gloom, when many begin to wonder if declines never halt, the appropriate abracadabra may be: "*They always did.*""

Bennett's book cannot save the billions of dollars wasted thus far on asbestos removal. But anyone reading it should learn that in science as well as on the stock market "*two and two still make four.*"

But it's going to take an informed public and media to make federal regulatory agencies accept that basic fact. Instead, the agencies have adopted a mathematical method of assessing dangers from such potential carcinogens as asbestos under which two and two make five or 10 or 23 or any other number the particular agency wants. The result has been a general perception that even the slightest exposure to a carcinogen is automatically life threatening, despite the fact that carcinogens are found in everything we eat, drink and breathe, even the sex hormones, progesterone and testosterone, which make new life possible. In the case of asbestos, that translated into the utterly absurd axiom, implicitly endorsed by the Environmental Protection Agency for years, that "one fiber can kill."

Yet everyone everywhere breathes asbestos with every breath because it is a mineral in the earth that breaks down by natural processes as well as in mining and milling, and is released as fibers into the air. God or nature put it there, and only God or nature can take it away. Still, EPA, in formulating asbestos regulations, completely ignored these basic scientific facts.

It's not entirely the agency's fault. Congress, in its infinite wisdom in writing environmental laws, has apparently assumed death can be outlawed.

However, the agency can be blamed for callously and cynically dumping responsibilities for dealing with asbestos on local government authorities without providing them with adequate guidance and assistance.

Indeed, William Ruckelshaus, former EPA administrator, admits in

Foreword by Reed Irvine

The Asbestos Racket that the theory behind the agency's original asbestos in schools rule was to "get the mothers to form a mob and storm the school committee."

That was and still is the official policy of the federal government, the deliberate subversion and undermining of the school boards which are perhaps the government bodies closest to the people as well as the guardians of our most important resource, our children and their future.

And, with the exception of Bennett's 1985 *Detroit News* series on asbestos and other environmental carcinogens, that damning fact has gone almost completely unreported in the media.

The federal government, along with the national media, particularly *The New Yorker* magazine, and the asbestos abatement industry, have hoodwinked, deceived and manipulated the American public into believing an almost nonexistent danger, asbestos, warrants a $150-$200 billion ripout campaign that places the lives of thousands of workers at needless risk.

And that's why, as chairman of Accuracy in Media, I believe *The Asbestos Racket* to be must-reading for anyone concerned about the environment, our schools, other public as well as private buildings, the conduct of the federal government and the performance of the media.

Bennett's book shows how legitimate concern over excessive exposure to asbestos among shipyard workers under wartime conditions has been perverted into a $150-$200 billion asbestos ripout racket. Even if all asbestos uses in the United States are banned, which the EPA wants to do by the year 2000, the net result might be 200 lives saved among building occupants.

Yet every scientist who has published peer-reviewed articles on asbestos has warned that the lives of tens of thousands of asbestos workers may be cut short by sloppy removals and disposals.

Further, as Bennett shows, as much as a trillion dollars in real estate values could be lost as a consequence of asbestos panic.

That is madness, mass madness. But as Charles MacKay observed 150 years ago, "Every age has its particular folly: some scheme, project or phantasy into which it plunges, spurred on by the love of gain, the necessity of excitement, or the mere force of imitation." That is as true now is it was in 1841, and modern environmentalism is the classic example in our times.

But it is well to remember that environmentalism is nothing new and has always had a sinister side. We tend to associate today's environmentalists with the poet William Wordsworth and his battles to keep "villas" and railroads out of the Lake District of England. The elitist nature of such campaigns, which essentially protect those who already have it made at the expense of those trying to move up the ladder, is deplorable, but certainly within the range of reasonable debate. That assumption, of course, may be rejected by, for example, the tens of thousands of workers expected to be thrown out of their jobs by the Clean Air Act of 1990.

But there is an absolutist strain in environmental thinking that can—and did—lead to the totalitarian excesses of 20th century.

The modern environmental movement began among youth organizations in Germany after the Napoleonic Wars. They dedicated themselves to the protection of the German race living in the "clean" countryside as opposed to the "polluted cities."

"It is no accident then that the Greens, as an organized political party, are most numerous in Germany, for it is possible to trace a thin line of continuity between the students of 1817," Paul Johnson, the well known historian wrote in a special Earth Day feature of *The Washington Times*, April 20, 1990, "and the much better organized youth movement just before World War I, with its camps, guitar playing and nature-worship."

Adolph Hitler inherited this phenomenon and turned it into his own Hitler Youth. He was a lifelong Green: vegetarian, non-smoker and non-drinker, dedicated to animals, especially endangered species—and, of course, an antivivisectionist.

As Johnson points out, environmentalism can be another example of religion-substitutes in the 20th century—Marxism, Fascism, Nazism, Maoism. In the United States and Great Britain, it takes the form of Puritan non-conformity, and involves some self-sacrifice. But as recompense, it offers the ineffable joy of self-righteous denunciation of others for "polluting."

"As agendas that appeal strongly to the religious-minded," Johnson concluded, "Green programs contain a powerful element of irrationality. Of course, they are usually couched in pseudo-scientific jargon, accompanied by impressive arrays of statistics, and endorsed by distinguished scientists and doctors. So, it is worth pointing out, were Nazi race policies. But the irrationality is always there and indeed is essential to their appeal."

Indeed, the current scientific understanding about cancer and the environment, a revolution in biology and biochemistry on a par with the overturn of Newtonian physics by Albert Einstein, has been virtually ignored by the mass media.

Distinguished scientists such as Dr. Bruce Ames, chairman of the department of biochemistry at the University of California at Berkeley, are showing that the rodent tests that the EPA has relied upon to determine which substances cause cancer are fundamentally flawed. Yet, despite this basic agreement among scientists that the results of such tests are useless, the voters of California were asked in a 1990 ballot initiative to ban growers from using a list of twenty to sixty pesticides. The voters rejected the measure by a substantial margin.

We are living in an era in which environmental efforts to return to an imaginary Garden of Eden free of carcinogens has produced a new

xii

Foreword by Reed Irvine

Inquisition. "Many of the risk assessment procedures used today are logically indistinguishable from those used by the Inquisition," as Dr. William C. Clark of the John F. Kennedy School at Harvard has observed in his monograph *Witches, Floods and Wonder Drugs: Historical Perspectives on Risk Management.*

As Clark observes, the Inquisition, like the environmental movement, enlisted experts "to prove explanations of the unknown and to mitigate its undesirable consequences." He points out that "creative and energetic efforts to create a witch-free world unearthed dangers in the most unlikely places; the rate of witch identification, assessment and evaluation soared."

As a direct consequence, Clark observed: "By the dawn of the Enlightenment, witches had virtually been eliminated from Europe and North America. Crop failures, disease, and general misfortune had not."

The costly efforts to identify and eliminate cancer-causing agents in the environment seem doomed to similar failure, since the scientific evidence increasingly shows that the targets of the EPA "witch hunts" are not the causes of significant cancer deaths in the country. The efforts of the EPA are not only wasting resources, but are often, as in the case of asbestos cleanups, counterproductive.

Many members of the public have long recognized many of the claims of the environmental lobby are based on unproven premises and hypotheses, which, taken far enough can lead to literal insanity. Everything the scientific community has learned about cancer recently can be summed up in the bumper sticker phrase, "Everything Causes Cancer."

There is no fundamental logical difference between the thinking of Nazis who believed the only "final solution" to the problems of the world was the eradication of the Jews, and environmentalists who believe only the eradication of asbestos, and chemical pesticides, can be the only "final solution" against cancer.

There are no "final solutions." There are only questions and rules of order. When we throw those out, whether in law, science, politics or journalism, we might as well go back to the howling jungle.

Mike Bennett is one of those journalists who understands that while we as individuals must always follow our own philosophical and political stars, our social and political ships are leaky and our captains subject to all the frailties of human beings.

The most devastating criticism is words that come back to haunt. In 1975, The Center for Responsive Law, headed by Ralph Nader, commissioned a book by Philip Boffey, later to join *The New York Times*, titled *The Brain Bank of America: An Inquiry Into the Politics of Science.* The book was a

xiii

The Asbestos Racket

classic example of the much vaunted advocacy journalism of the time. The book was designed, in Nader's words, to develop and promote "scientists-as-citizens" in regulatory agencies: "The number of full-time public interest scientists must proliferate in the coming years if science for the betterment of mankind is to be loosed from the irons of the parochial, if not avaricious demands of the corporate state."

Well, the scientists-as-citizens did get turned loose in EPA and other federal agencies. And the result has been an asbestos racket that is costing the country as much as the S&L bailout in direct removal costs, $150-$200 billion, and as much as $1 trillion—an amount equivalent to the national debt—in depreciated property values, not to mention the lives of perhaps thousands of asbestos abatement workers unnecessarily shortened by poorly performed asbestos removal projects.

To anyone, including the most dedicated environmentalists, after reading *The Asbestos Racket*, these words of Philip Boffey must resonate with tragic irony down the years:

> "Ours is a society that believes in expertism, that constantly genuflects before the presumed wisdom of experts. Advisory groups and individuals are routinely called upon to counsel everyone from the President to corporation heads to the local authorities in charge of laying sewer pipe. The public tends to assume that these expert advisors dispense some sort of objective truth, the 'Right' answer to the problem under consideration. But such implicit trust is misplaced. There are relatively few public policy questions whose answers are misplaced. In almost every case, an element of informed judgement is required, and what comes out strutting as 'objective' wisdom is actually the subjective wisdom of those who prepared the advice. Unfortunately, those expert advisers can be just as biased and pigheaded as you and me, and they can be just as foolishly wrongheaded as we often are."

Environmentalism, like the Inquisition and Communism, has denied the basic rules of fairness and civility that have traditionally governed science, in the form of peer review; journalism, in its basic respect for all sides of a story; and politics, in its decent regard for the rights of all. For Environmentalism, like Communism or Fascism, exalts ends over means, results over process.

And just as with any other fanatic faith, the result is not heaven, but hell on earth.

Life is hard on rayform movements in America;
they either turn into corpyations or rackets.
—Mr. Dooley

1 Asbestos Hysteria

"Cancer is now in the service of a simplistic view of the world that can turn paranoid," warned Susan Sontag, the literary and social critic, in *Illness as Metaphor* almost 15 years ago.

Her prophecy has come true—and nowhere has it been more clearly demonstrated than with asbestos, a natural mineral which has been around since the beginning of time. Since it is a mineral found in the air, water, and soil, people have been eating, drinking, and breathing asbestos since Adam and Eve walked the Garden of Eden.

But in modern America asbestos has been cast in the role of the snake in the Garden of Eden, a symbol of all that is presumably wrong in modern society. The result has been mass hysteria for profit. The profit is not just for asbestos abatement companies, but also for empire-building bureaucrats, fund-raising environmental groups, and sensation-seeking media. It is institutionalized environmental and legal hysteria which proves that we are still only a step away from the jungle or the Inquisition or delusions of racial supremacy.

For the theories on which asbestos hysteria are based have no more scientific validity than those supporting any of the other mass hysterias which have gripped the human race from the days of Torquemada's witch hunts to the mass madness of Nazism or Communism.

Science has only recently had any influence on asbestos policy. William K. Reilly, administrator of the Environmental Protection Agency, made the first bureaucratic bow to science in a June, 1990, speech titled, "Asbestos, Sound Science and Public Perception:

1

Why We Need a New Approach to Risk."

Reilly, a former advocate for environmental groups, prefaced his speech with an anecdote from his confirmation hearings. He had spent an hour with Sen. Daniel Patrick Moynihan (D. - N.Y.), who before his election to the Senate was perhaps the best recognized—if controversial—political science scholar in the country.

"He sat me in a very nice Windsor chair," Reilly recalled, "about which he said, 'This is a Republican chair. . . . This is appropriate, I think for the new EPA administrator to sit in.'"

EPA was created under an executive order of former President Richard Nixon back in 1970, and unlike any other permanent agency or department of federal government has no legal authority based on a fundamental charter issued by Congress after appropriate debate and agreement. Instead, it reports to a hodgepodge of some 82 Congressional committees and subcommittees, all with their own agendas and politically influential—and generous—lobbyists.

As Reilly recalled, Moynihan, "perching his little half-reading glasses on his nose said, 'Above all, above all, do not allow yourself to become transported by middle class enthusiasms.'"

"Well, I assured him that wouldn't happen, and later I came to a conclusion about what he meant. What he meant, I decided, was 'Respect sound science; don't be swayed by the passions of the moment.'"

But up until then that simply hadn't been the case at EPA.

Why did that encounter happen? In due course I will offer an explanation and suggest that it may mark a fundamental turning point in the politics of environmentalism.

Reality has intervened.

Even professional environmentalists—including Reilly, who had been president of the Conservation Foundation—have been forced, if only by political necessity, to recognize the overriding value of science. A fundamental conflict, which had been dubbed "the lawyers vs. the scientists" syndrome, in which the lawyers and policy-making managers always won and the scientists always lost, was being dominated by politicians rather than scientists.

And it is a conflict whose roots go back to 1970 when EPA was established by Nixon to satisfy "middle-class enthusiasms."

"We are all Keynesians now"—and environmentalists as well, Nixon proclaimed in successfully seeking to divert support from middle-class enthusiasts drawn to the political campaigns of Sen. Edmund Muskie, author of early environmental legislation, and of Sen. George McGovern.

The trouble was that some people, including a fellow Californian by the name of Ronald Reagan, feared the economic effects —and no one really understood what environmentalism was all about.

Unlike the vast majority of political and social movements, environmentalism seems to have slim intellectual underpinnings, except for Rachel Carson's *Silent Spring,* and that was narrowly focused on the pesticide DDT. Of course, the conservation movement had been around for more than 100 years. Still, because it was divided on internal sectarian lines, exemplified by utopian "nature worshippers" such as John Muir, founder of the Sierra Club; utilitarian "wise users" like Gifford Pinchot, first Chief of the U. S. Forest Service; and genteel greenery advocates such as Frederick Law Olmsted, designer of New York's Central Park, Washington, D.C.'s Mall, and Boston's "Green Necklace," it was hardly a mass movement.

Environmental consciousness seemed to come out of nowhere—in fact, out of the television tube, which spawned late twentieth century images in living color of an apparently fragile "Spaceship Earth" or of ecological disaster from an oil spill in Santa Barbara, California (yet natural oil spills in the same channel were recorded as far back as the 17th century according to a 1970 article in *Science*).

Environmentalism's most fervent supporters, as William Manchester wrote in *The Glory and the Dream: A Narrative History of America, 1932-1972,* "were flagrantly inconsistent. Although vehemently opposed to pollution of the environment, for example, they enthusiastically supported by their patronage another form of pollution: the junk food industry."

It is worthwhile to examine briefly the notion of pollution itself. The word "pollution," which prescribed some sensible dietary restrictions in both the Jewish and Islamic traditions in the Middle and Far East, went undescribed by environmental enthusiasts. The word *pollution* was imputed as having a scientific meaning, but, at root, it simply re-echoed, in modern form, the old cry of "moral"—not hygienic—uncleanliness.

An entire "Children's Crusade" was led by perhaps the most humorless—albeit secular—preacher in the history of this country except for Cotton Mather: Ralph Nader. According to Manchester, Nader's Lebanese immigrant father never let the customers in his restaurant and bakery "eat in peace. [He] was always lecturing them about the wrongs, the inequities, the injustices of the system. Like many immigrants, he was a more ardent Democrat than the natives. He went on about the crimes of the Interests and was forever threatening to sue them. In time nearly everyone there tuned him out, with one exception: his youngest son Ralph."

And Ralph, already Puritan, became a lawyer. As Manchester wrote, "he foreswore the reading of novels; they were a waste of time. So were movies; he would limit himself to two a year. He scorned plays, tobacco, alcohol, girls, and parties."

After Harvard Law School, he got himself a job as a fifty dollar-a-day consultant to none other than Daniel Patrick Moynihan, then Assistant Secretary of Labor for Policy. From there through publication of his book *Unsafe At Any Speed: The Designed-In Dangers of the American Automobile*—a now-repudiated attack on the first major subcompact car—Nader was to become the nation's leading consumer expert.

What a "consumer expert" was, certainly, few if any knew. In fact, one observer said of Nader: "Ralph is not a consumer champion. He is just plain against consumption." And, in the period of the 1960s, everyone who was anyone was a "middle-class enthusiast."

It was an age when the zeitgeist was "the spirit of overkill," when the famed philosopher Bertrand Russell said that American President John F. Kennedy and British Prime Minister Harold MacMillan were "worse than Hitler." And it was an "age of publicity" in which outrage was the common currency of communication. Susan Sontag could write in *The Partisan Review*, in her revulsion against the Vietnam War, "the white race is the cancer of history." Only later, after experiencing cancer personally was she to write in *Illness As Metaphor* that the use of cancer as a metaphor for anything is "morally impermissible."

Nader, and most environmentalists, were above all, in Manchester's phrase, "linear thinkers" whose thought processes flowed only in one direction. If "A" was correct everything from "B" to "Z" was equally correct with no allowance for course change, interventions, ambiguities, events, hesitations, questions, quarks, uncertainties or ironies.

Presumably every slope was slippery, and when they were paved by capitalists led inevitably and ineluctably to hell. The best educated generation in history—at least terms of years of college attended—had apparently taken over the ship of state, morally if not yet politically, and with all the force of their moral conviction were driving straight forward with no allowance for sea changes. Henceforth, there would be no tacking to lee or starboard or coming about. Only conviction, not seamanship, counted.

And in the environment, although certainly not in politics, it was possible to sustain that illusion for years. In the first place, people literally didn't know what they were talking about.

The vast majority of college-educated Americans were at least one or two generations removed from the farm and the ranch and the woods and the practical realities of wresting a living from nature—the environment. At the same time, middle-class suburbanites, encouraged by park systems, movies and, increasingly, television, were nature *worshippers*, in a way Europeans with cultural memories of war, famine, and recultivation could never be.

More Americans knew the words to *America the Beautiful*—
"from purple mountain majesties / across the fruited plains"—during
the '50s and '60s than to the *The Star-Spangled Banner*. And in the
1960s they were more likely to be singing "This land is my land, this
land is your land. . . ."

Even the word *environment* had been usually employed to
describe primarily urban social conditions—"Don't send him, away,
judge; he's basically a good kid, but he came up in a bad environment." The word *environment* in fact, was almost never used to
describe a state of nature until the '60s.

There was a relatively obscure scientific discipline called
"ecology" around, which had been developed in Germany in the late
19th century. Its principal founder, a zoologist named Ernst Haeckel,
defined ecology as "the study of the interrelationships of an organism
to the sum total of both its inorganic and organic life conditions;
including, above all, its active and submissive relations with those
animals and plants with which it comes directly or indirectly in
contact—in short, ecology is the study of all the complex interrelations
referred to by [Charles] Darwin as the conditions for the struggle for
existence."

Ecology, therefore, was rooted in the central scientific principle of 19th and 20th century biology: evolution. Haeckel himself
was a strident Darwinian. And within that framework, there is no
constant except change. But ecology, although it encompasses so
many factors, interrelationships, and uncertainty concepts that it
would bewilder the mind of a Shakespeare or Aristotle or Plato or
Immanuel Kant, all accustomed to images, symbols, and ironies, is
still a science.

"Ecology *is* a science," as George Claus and Karen Bolander
wrote in *Ecological Sanity*. "It is exceedingly complex; nevertheless
it is a discipline, the problems of which have to be attacked by
scientific methods within the framework of our scientific causality. It
is not a *Weltanschauung* (world-view) nor a pantheistic religion."

That central point, that basic mental construct, got lost in the
shift during the 1960s from ecology as a science to environmentalism

as a *Weltanschauung* or a pantheistic religion. Also metaphorically lost was any social equivalent of the "uncertainty principle" first enunciated by Werner Heisenberg—the quantum principle which states the impossibility of determining more than one observable quantity at a time, such as finding both the position and velocity of an electron.

Translated into ordinary language, the most common expression would probably be, "You can't see the woods for the trees." Anyone who looks only straight ahead will inevitably be blind-sided. The helmsman who focuses only on the North Star or Southern Cross will inevitably end up on the rocks. The lawyer who argues exclusively on the basis of precedent will be tripped up by new evidence. And scientists have to be constantly aware, no matter how carefully or meticulously their experiments are conducted, that another experiment, using exactly the same protocol can, with one simple change of procedure or perception, can come up with entirely different results.

Reality always confounds linear thinking.

Columbus was absolutely right in thinking that by sailing due west he would eventually reach China. He was right—but two continents stood in his way. For more than 200 years, scientists correctly assumed that most disease was caused by inadequate hygiene; it took almost fifty years to discover the most feared and primarily "American" disease of the early and mid-20th century—polio—was engendered by American baths and toilets that literally washed away a natural immunity which could only be restored by immunization. And, of course, smallpox, once the most prevalent and now the only extinct disease on earth, was wiped out by literally giving people a light case through inoculation.

Facts have a sneaky way of creeping up on the best made plans of mice and men—including those who experiment upon and are experimented upon in the search for the answer to cancer, today the most feared disease in America.

Americans are guaranteed the right to the pursuit of happiness—but rarely, if ever, achieve it on a straight course.

God is not a linear thinker.

And that is why the asbestos experience can be so important, for it permits us to see, as an historic example, how we can be both as smart and as stupid as the people who emerged from the Dark Ages to build the great cathedrals of Europe and simultaneously debate how many angels can dance on the head of a pin.

In the second millennium, as well as the first, wisdom starts by defining our premises, sorting out our terms.

The asbestos experience provides a classic example of why we all, as fallible and ambitious and greedy and well-intentioned human beings, fail to understand how we think, or if, in fact, we think.

It is a study not just in science or in politics, but also in culture, logic, and irony, and in epistemology, the basic discipline of philosophy that investigates the origin, nature, methods and *limits* of knowledge.

For it demonstrates that knowledge is based on a continuum, a range, a spectrum, lines that inevitably curve back upon themselves.

Most people instinctively understand that fact in their own lives. They don't think entirely in terms of their own specialized knowledge. They may be lawyers or laborers, scientists or salesmen, police officers or dentists, but they share a real rather than a theoretical environment. It is an environment in which the lines always curve.

But Washington doesn't operate that way.

It is linear, particularly on issues and problems which are defined by the news media, rather than personal perception, by pollsters rather than people.

Power may originate from politics beyond the Beltway. But once it is concentrated in Washington, the seat of power, it is a zap ray back out. "Middle-class enthusiasms" are always tempered by reality outside Washington, because they always have to be paid for in one way or another. Theory always runs up against reality.

But in Washington, theory is always the building block of empire and only backs off when the entire empire itself is threatened, as when the body bags started coming back from Vietnam.

And now the body bags are coming back from what Richard Nixon declared, in founding EPA a generation ago, as the "War on Cancer."

"We are destroying the village in order to save it," as a possibly apocryphal Army officer was quoted by Peter Arnett of Associated Press in Vietnam. But now, as then, there is "no light at the end of the tunnel," only misplaced visions and wasted efforts, for better or worse.

And, in the war on cancer, EPA's asbestos policies are now being seen as an ironic counter-image of the middle-class enthusiasms which led to the revulsions against the Vietnam War. Only now we're physically "wasting" our own schools to the tune of $200 billion at a time when resources are sorely needed to educate our children, not just as a social imperative, but also as an economic necessity in an increasingly competitive world.

Linear thinking has once again run amok and is threatening to imperil the very people we are presumably trying to save. For, as Reilly said in his speech, "An excellent example of a clash between real risks and public perceptions is the current controversy over asbestos in the nation's schools and public buildings.

"As a conservationist and as a lawyer, I have a good deal of experience with what has been called the 'law of unexpected consequences.'

"Our experience with asbestos is a good example of that law in action."

Unfortunately, the only attention Reilly's speech received in the mass media was summarized in an op-ed piece in *The Washington Times* on August 15, 1990—and I wrote it. The rest of the media, having been thoroughly brainwashed by the environmental movement and EPA's public relations apparatus, ignored the story.

But the story—and a whole new mental and policy-making framework—is now being constructed, fueled by complaints from school officials, parents, teachers and maintenance workers. That cascading concern has already been translated into action in the form of proposed legislation to repeal the Asbestos Hazard Emergency Response Act (AHERA), a repeal introduced by Sen. Malcolm Wallop (R. - Wy.) and endorsed by many other members of Congress including Sen. Max Baucus (D. - Mt.) and Sen. John Chafee (R. - R.I.), respectively chairman and ranking minority member of the Senate Environmental and Public Works Committee.

The specific language of the bill was important, but the overriding concern expressed by Sen. Wallop was far greater. For he saw in asbestos a paradigm, a model, a standard, a touchstone, of what Sen. Moynihan called "middle-class enthusiasms."

The lines are curving.

In addressing the asbestos problem, a way may be found to develop a rational, humane environmental world view, a comprehensive way of looking at the world and man's role in it—without using the passion of the moment as a way to flagellate the sins, real or perceived, of the past.

As Sen. Wallop said in introducing the legislation, "The Congress indeed has the responsibility to legislate. But it does not always have the ability to legislate wisely. Too often, we are a reactive body, passing legislation in haste when we are confronted with a crisis or the perception of a crisis. And here is one of those times.

"We act without knowledge, without reflection, and it is a poor standard, a terrible example, for what we call the world's greatest deliberative body."

As Wallop pointed out, results can be terribly disruptive in terms of education as well as economically wasteful. In Casper, Wyoming, for example, one of the two high schools in the town had to be closed down at a cost of $4 million for totally unnecessary asbestos removal and the students and teachers placed on a double shift, from 7 a.m. to 1 p.m. and 2 p.m. to 7 p.m.

"Fortunately, Congress can undo its mistakes, once it is willing to admit them," Wallop said on March 7, 1990. "That is the purpose of the legislation I am introducing today."

And that is the purpose of this book.

Almost anything carried to its logical extreme can become depressing, if not carcinogenic.
—Ursula LeGuinn

2 EPA'S Asbestos Farce

EPA Administrator William Reilly in 1990 came close to admitting that his organization is responsible for the greatest environmental fraud of our era. In a speech before the American Enterprise Institute, Mr. Reilly admitted that much of the EPA's work on asbestos at best was riddled with errors and had proved unnecessarily expensive in terms of lives and money.

"The mere presence of asbestos poses no risk to human health," he said, "only when asbestos fibers are released into the air and breathed into the lungs do they become a human health risk." He then contended that EPA has been arguing against the unnecessary removal of asbestos, since removal efforts "may actually pose a greater health risk than simply leaving them alone."

The speech was fascinating because Mr. Reilly subtly reversed his agency's historic position on asbestos and then misrepresented EPA efforts. In fact, the EPA has only reluctantly, and under great pressure from the scientific community, admitted that removal causes more harm than good. EPA hasn't fundamentally changed its policies.

The agency's first "authoritative" guidance book described removal as "the only final solution" to asbestos, and, until recently the EPA had imposed this "solution" on businesses and citizens alike.

Asbestos policy and its quiet reversal by the EPA, typifies the environmental lobby's penchant for focusing on only one side of the economics-environmental equation. Forces eager to outlaw asbestos overlooked the fact that asbestos has probably saved tens of thousands of lives each year as a fire retardant. After a fire in the Coconut Grove night club in Boston in 1942 killed 490 people in less than an

13

hour, almost every state and community on the nation passed laws requiring the installation of asbestos in private and public buildings–and the material is still considered far more effective than the available substitutes, i.e. rock and glass wool. Environmentalists also conveniently overlooked the fact that asbestos comes in two basic forms, chrysotile and amphiboles. Chrysotile, which is the material found in 95% of all buildings, is far less dangerous than the amphibole varieties.

The EPA in 1982 accepted the hysterical claims of professional environmentalists "that there can be no safe level of exposure to a carcinogen"–or in the case of asbestos, "one fiber can kill." The agency also accepted a scientifically discredited four-year-old study predicting asbestos as a low level carcinogen would cause as many as 40,000 "excess deaths" per year. The study was based on the experience of World War II shipyard workers who were faced with extremely high levels, with the results extrapolated to the general public.

The EPA did so despite the fact that Dr. Malcolm Ross of the U.S. Geological Survey had clearly demonstrated the actual number of annual asbestos deaths was no more than 520 at the highest point–and was falling sharply as the shipyard workers died off of natural causes. The EPA also ignored its own scientific review of the study used to justify a ban on asbestos in schools. The EPA's own scientific panel denounced the study as "unconvincing," "greatly overestimated," "scientifically unappealing," and "absurd."

The agency also ignored Sir Richard Doll of Oxford University, the world's leading epidemioligist, who compared the risk from asbestos to building occupants as comparable to smoking one-half a cigarette in a lifetime.

The EPA's original policy on asbestos was designed, as admitted by former Administrator William Ruckelshaus, to "get the mothers to form a vigilante mob to storm the school committee," because otherwise "the federal government would have to pay for it [removal] and the cost would be astronomical."

The agency failed to anticipate the reaction of businesses and schools threatened by an asbestos policy that ordered them to remove the substance from ceilings, walls, even floors—no matter what the cost. But business and schools did respond, and no group was more vocal than the U.S. Catholic Conference, whose schools were literally being driven out of business by EPA asbestos removal policies.

Reilly admitted in his 1990 speech was that it was largely prompted by "recent meetings I have held with school officials—including a delegation of the U.S. Catholic Conference—and members of Congress." EPA's new found passion for science, in other words, is not entirely non-political.

EPA's asbestos policies are raising the "question of the very survival of parochial schools," said Archbishop Theodore E. McCarrick of Newark, N.J., co-chairman of the U.S. Catholic Conference delegation that met with Reilly. The cost of asbestos removal, estimated at $150 billion to $200 billion—and upwards of $1 trillion in lost property values, four times the savings and loan bailout—is enough to make any politician think twice.

This includes Mr. Reilly, who admitted in 1989 that asbestos was responsible, at worst, for 15 deaths a year and has received some pointed political advice from men such as Sen. Moynihan who told the administrator to "avoid middle-class enthusiasms." Mr. Reilly's change of heart may also stem from Sen. Malcolm Wallop's bill to repeal the existing asbestos law.

Yet despite the administrator's seeming shift in philosophy, the EPA hasn't formally renounced its asbestos policy. Thousands of businesses remain at financial risk bordering on bankruptcy, and tens of thousands of people have been exposed to unnecessary health risks. The two senators, Moynihan and Wallop should demand hearings to explore the EPA's political hypocrisy and to discover why the agency hasn't fully abandoned its ridiculous policies. This much is certain: Mr. Reilly ought to drop the curtain on this tragic and absurd environmental farce.

The primary considerations to be borne in mind are two key scientific facts:

1. The only clearly established cases of asbestos-related disease occurred almost exclusively among workers exposed under wartime conditions to concentrations 10,000, 100,000 and one million times more than those allowed by current occupational standards and regulations.

2. There is no evidence whatsoever that asbestos in buildings has caused a single case of cancer or any other asbestos-related disease.

But within the linear mindset cultivated by the environmentalists almost no one was listening to the scientists, at least until early 1990.

However, the lines—the perceptual lines—were beginning to bend. The light was bouncing back.

"Congress passed a law banning asbestos, and schools have spent billions removing it," an editorial in *The Wall Street Journal* reported on February 6, 1990. Scientists are now calling this episode a "panic, and suggesting the money was wasted."

The Journal was commenting on a "major study, just published in *Science* [Jan. 19, 1990], which confirms the conclusions of a half-dozen other studies of asbestos, including a recent one by Harvard's Energy and Environmental Policy Center, which reached essentially the same conclusion."

The Journal was only half right. There have been dozens, if not hundreds of studies which have reached exactly the same conclusion. And many of them are ten, even twenty years old.

All came to essentially the same conclusions as the authors of the *Science* article cited by the *Journal.* The article was written by Brooke D. Mossman of the Department of Pathology at the University of Vermont; Jacques Bignon of the Unite Recherche sur la Biopatholgie et la Toxicologie Pulmonaire et Renale, France; Morton Corn of the Division of Environmental Health, Johns Hopkins University; A. Seaton, Institute of Occupational Medicine, Scotland; and J.

Bernard L. Gee of the Department of Internal Medicine, Yale University.

Their article, unlike those of their predecessors, received more media attention, not only in the *Journal* but also *The New York Times* and other newspapers because it was bluntly labelled as a public policy as well as a scientific paper: "Asbestos: Scientific Developments and Implications for Public Policy."

Nevertheless, the article, because it has received so much media attention, has marked a major breakthrough in public perception about the real and relative risks associated with asbestos. As such, it is worth quoting at length.

> Asbestos engenders both fear and panic in U.S. society. Observation that asbestos-containing materials (ACM) have been used in schools, buildings and hospitals and the Asbestos Hazard Emergency Response Act (AHERA), a mandate from the Environmental Protection Agency (EPA) that requires inspection of the nation's public and private schools, have resulted in an explosive growth of asbestos identification and removal companies. By EPA estimates, extension of EPA requirements to approximately 733,000 public and commercial buildings containing asbestos will cost $53 billion, discounted at 10 percent over 30 years. Because of uncertainties regarding the amount of asbestos and its condition in these buildings, estimates for removal of asbestos are as high as $100 to $150 billion. . . .
>
> "Asbestos" is a broad commercial term for a group of naturally occurring hydrated silicates that crystallize in a fibrous habit. . . . The family of asbestos minerals can be divided into serpentine and amphibole fibers.
>
> Chrysotile, which accounts for over 90% of the world's production of asbestos, is the most common fibrous serpentine, whereas the amphiboles [are] a chemically diverse group of less important fibers. . . .

The rod-like amphiboles appear to penetrate the peripheral lung more readily than chrysotile fibers, which are curly, can occur in bundles and can be intercepted at airway bifurcations. . . .

The available experimental and epidemiological data indicate that both fiber type and size are important determinants of the pathogenicity of asbestos. Although asbestos has caused disease in the workplace and such occurrence has resulted in calls for regulation to protect workers, recent epidemiologic data are concordant with the suggestion that exposure to chrysotile at current occupational standards does not increase the risk of asbestos-associated disease. Unlike most other countries, particularly in the European Community, which have more stringent requirements for regulation and importation of amphiboles (mined primarily in South Africa), federal policy in the United States does not differentiate between different types of asbestos.

Does airborne asbestos present a risk to the health of individuals in schools and other buildings? The available data do not indicate that asbestos-related malignancies or functional impairment will occur as a result of exposure to most airborne concentrations of asbestos in buildings. . . .

First and foremost, the levels of airborne asbestos in buildings, even with damaged ACM. . .are several thousandfold lower [than] the permissible exposure of 0.2 fibers per cubic centimeter of air in the U.S. workplace.

[Second,] With few exceptions, the type of asbestos fiber predominantly in buildings is chrysotile. . . . [T]he biologic effects should be should considered individually for each fiber type.

The authors warn that the mere presence of asbestos in a building literally means nothing. It is only dangerous when airborne in large quantities, and that determination requires "sophisticated technology such as transmission electron microscopy, x-ray diffraction or energy dispersive x-ray spectroscopy."

The public policy conclusions of the authors logically follow from those givens:

> The AHERA ruling of 1986 brought asbestos to the attention of the U.S. public and instilled fears in their parents that their children would contract asbestos-related malignancies because of high levels of airborne asbestos fibers in schools. Panic has been fueled by unsupported concepts such as the "one fiber theory," which maintains that one fiber of inhaled asbestos will cause cancer.
>
> As a result of public pressure, asbestos often is removed haphazardly from schools and public buildings even though most damaged ACM is in boiler rooms and other areas which are inaccessible to students or residents. The removal of previously undamaged or encapsulated asbestos can lead to increases in airborne concentrations of fibers in buildings sometimes for months afterwards, and can result in problems with safe removal and disposal
>
> Asbestos abatement also has led to the exposure of a large new cohort of relatively young asbestos workers. While these people should be protected by careful regulation of the circumstances of removal, they are often exposed under suboptimal working conditions.
>
> As a result of the AHERA ruling, public and private schools are required to inspect for asbestos and inform parents if ACM are present. Although the law does not require or set standards for the removal of asbestos, schools, often with little expert advice, must submit a management plan detailing how they will deal with damaged asbestos and can be fined a maximum of $5,000 per day for lack of compliance to deadlines.
>
> The EPA has recommended bulk sampling of ACM to determine the presence of asbestos and visual inspection to determine the course of action, rather than measurement of

airborne levels of fibers—data that are far more important in determining the need, if any, for removal of ACM.

The available data and comparative risk assessments indicate that chrysotile asbestos, the type of fiber found predominantly in U.S. schools and buildings, is not a health risk in the nonconventional environment. Clearly, the asbestos panic in the U.S. must be curtailed, especially because unwarranted and poorly controlled asbestos abatement results in unnecessary risks to young removal workers who may develop asbestos-related cancers in later decades. The extensive removal of asbestos has occurred less frequently in Europe.

Prevention (especially in adolescents) of tobacco smoking, the principal cause of lung cancer in the general population, is both a more promising and rational approach to eliminating lung tumors than asbestos abatement. Even acknowledging that brief, intense exposure to asbestos might occur in custodians and service workers in buildings with severely damaged ACM, worker education and building maintenance will prove far more effective in risk prevention.

But, if the *Science* article is the leading voice in the chorus, it is far from the only one. Other articles and most scientific authorities in the field have come to the same conclusions.

The likelihood that a single case of cancer could develop among building occupants exposed to asbestos, used as a insulating, fire protecting, and acoustical material is "probably close to zero." Those are the words of Dr. Tee L. Guidotti, professor of occupational medicine at the University of Alberta in Canada, based on a study requested by a union representing 2,000 workers in a 15-year-old Edmonton office building.

Following an intensive survey by a team of occupational experts, Guidotti concluded in an article printed in the July-August, 1988, issue of *The Canadian Journal of Public Health*, "These levels of excess risk are so low (probably much less than 3 in 10,000) that they would never be detectable in a surveillance study of this population. The level of risk reasonably could be compared, assuming a lifetime at risk of 70 years, to the probability that someone among the 2,000 occupants would be struck dead by lightning, a highly unlikely occurrence."

The risk for nonsmoking asbestos workers who have been exposed to much higher concentrations of the mineral are not much greater. Even the 1950s and '60s studies of Dr. Irving J. Selikoff of Mt. Sinai School of Medicine, which EPA relied upon in formulating a 1989 asbestos ban, estimated cigarette smoking multiplied risks 80 times over. Those findings were further confirmed by a study of asbestos workers published in the book *Asbestos-Related Malignancy*, published in 1987.

Indeed, "it remains uncertain whether any type of asbestos acting alone can cause lung cancer in non-smokers," concludes a comprehensive review of the scientific literature by two of the authors of the *Science* article, Drs. Mossman of the University of Vermont and Gee of Yale University. Their review of the literature appeared in the June, 1989 issue of the *New England Journal of Medicine*.

The study concluded, as have dozens of others by universally acknowledged scientific authorities all over the world:

"In the absence of epidemiological evidence or estimations of risk that indicate that the health risks from asbestos are large enough to justify the high expenditure of public funds, one must question the unprecedented expenses on the order of $100 billion to $150 billion that could result from asbestos abatement."

Death from smoking is 21,900 times more likely than exposure to asbestos in buildings, according to the Energy and Environmental Policy Center at the John F. Kennedy School at Harvard. Comparable risks include death among motor vehicle passengers, 1,800 to 1; frequent plane passengers, 1,600 to 1; coal mining, 441; pedestrians, 290; living with a smoker, 200; cycling, 75; drinking water in New Orleans or Miami, 3; and lightning, 3.

The Center report, based on an international symposium held in December, 1988, concluded:

"● Recent measurements indicate that the average levels of airborne asbestos in buildings containing asbestos materials are extremely low. 'Fiber phobia' among the general public, therefore, is out of proportion to the existing health risk from building exposures.

"● . . .The health risk posed by in-place asbestos is very small, both in absolute and relative terms, and is far less than most other commonly experienced environmental hazards.

"● . . .Concern should focus in maintenance and utility service personnel whose occupations place them in close physical contact with asbestos containing material that may be disrupted.

"● . . .More recent scientific evidence indicates that removal of asbestos, if improperly done, may actually increase health risk not only to removal workers but also to the building occupants."

The report also confirmed another key point about the health effects of asbestos: The adverse health effects of chrysotile or "white" asbestos, the type found in 95% of all buildings, are minimal by comparison to the types implicated in Selikoff's original studies of wartime workers. Amosite or "brown" asbestos and crocidolite or "blue" asbestos—the so called amphibole varieties—are mined in Australia and South Africa.

The two types were only used in any substantial quantities in the United States during World War II. Supplies of chrysotile, mined primarily in Canada but also in the United States, were in short supply during the war. The amosite and crocidolite were used primarily in ship construction, which employed most of the workers studied by Selikoff.

The Harvard study and dozens of other authorities have

concluded:

"'Asbestos' is not a single mineral. It is a term used to refer to a number of fibrous inorganic minerals. The most common of these is chrysoltile, the predominant fiber type of all asbestos used in the United States. . . . Chrysotile fibers, because of their instability in tissues and other characteristics, present a substantially lower risk of malignancies than amphiboles."

Chrysotile fibers are short, thin, and curly, and, therefore, are normally dissolved or expelled from the lungs soon after being breathed in. Amphiboles, by contrast, are much more likely, because of their greater length—more than 5 microns—to lodge in the lungs. Therefore, years later, especially in combination with cigarette smoking, disease is much more likely to occur.

The evidence against the amphiboles, particularly crocidolite, which was used in considerably greater quantities than amosite, is incontrovertible. The most recent evidence comes from a study conducted by a team of researchers at the Dana Farber Cancer Research Institute in Boston. The findings were printed in the November, 1989 issue of the *New England Journal of Medicine.*

The 33 men in the group studied made the first filter touted by the cigarette industry as making "coffin nails" safe. Between 1951 and 1957, the 33 worked for the Lorillard Tobacco company making "Micronite" filters. The average term of employment was 1.7 years.

The presumably "secret" element in the "Micronite" filters was crocidolite or blue asbestos.

Only eight of the 33 could be expected, as a matter of statistical probability, to die in the ensuing years.

Twenty eight have died, 325% more than anticipated.

All but one of the 28 died of asbestos-related disease, 11 from lung cancer; five from mesothelimoma, a rare disease of the lung cavity almost invariably attributed to amphibole asbestos; five from asbestosis; three from other forms of cancer; and two from respiratory disease.

If those workplace exposure figures were extrapolated on a mathematical model similar to the one used by EPA to calculate risks to the general public, almost 17.9% percent of the people in the country would already be dead from asbestos-related disease.

But, of course, they are not, because of two well known and very clearly understood scientific principles:

Health response and effects are determined by
1. Toxicity, and
2. Dose, or
Toxicity x dose = effect.

To illustrate with two well-known examples: Arsenic, a highly potent poison, is found in some shellfish including lobsters. But the quantities of arsenic found in lobster are minute. Consequently, there are no recorded cases of normal human beings dying from eating lobsters. Similarly, some mushrooms are hallucinogenic. But in most such mushrooms, small portions will not induce hallucinations. Toxicity x dose = effect.

EPA's asbestos record on the contrary, would suggest that hallucinogens are regularly ingested in large quantities at EPA. Certainly, basic scientific and logical principles are not understood there.

For, even in EPA's worst-case scenarios, only one of 100,000 persons exposed to chrysotile (white) asbestos at concentrations of 0.001 fibers per milliliter of air (f/ml) for 10 years can be expected to develop any kind of disease under the agency's mathematical models. Yet the average concentration of asbestos in school and other buildings is about 0.0007 f/ml. About as much is found in outside air. And fibers longer than 5 microns were even lower, 0.0008 fibers, both inside and outside.

Even EPA's own worst case findings in a sampling of 43 federal buildings found levels no higher than outside air.

"At these levels," the Harvard report concluded, "risks are less than one in a million." In other words, only 200 to 250 persons can be expected to develop asbestos related disease by the end of the century.

The Health Effects Institute (HEI) of Cambridge, Massachusetts, which sponsored the Harvard symposium, is developing what should be a "definitive. . .assessment of airborne asbestos levels in buildings containing asbestos." That report, which is expected late in 1991 or early in 1992, should, in an ideal world, finally close the issue.

The Institute was established with the active support of William Ruckelshaus in his first term as EPA administrator, under the Clean Air Act, one of the earliest environmental laws. The first mission of HEI, which is funded by 50/50 contributions from industry and the federal government, was to monitor potentially harmful emissions from automobiles.

The Institute's integrity and credibility are unmatched, partially because its president is Archibald Cox, famed for his role as the special prosecutor in the Watergate investigation. Ruckelshaus, of course, was fired as assistant attorney general by former president Richard M. Nixon for refusing, along with Attorney General Elliott Richardson, to fire Cox as the special prosecutor.

Moreover, HEI was brought into the asbestos issue by former Rep. Edward Boland (D. - Mass.), as his final action as the chairman of EPA's appropriations committee. He saw to it, with EPA's agreement, that $2 million was appropriated to finance the "definitive" HEI study. Another $2 million on a "50/50 basis" is coming from grants from private interests, including current and former asbestos product manufacturers, realtors, developers, building owners and managers, mortgage bankers, the insurance industry, and labor unions.

Given HEI's credibility and EPA's own assurance that it would accept the findings of the final report, it would seem logical and prudent for the agency to defer any final decision on a total ban. The ban was originally proposed in 1985. In the interim, EPA had sharply scaled down its estimates of deaths due to asbestos from 1,800 by the end of the century to 500.

But, EPA nevertheless, went ahead with the announced ban on July 6, 1989, despite the fact that the interim HEI report was being widely circulated in the agency and elsewhere. However, if imitation is the sincerest form of flattery, EPA Administrator Reilly paid HEI an extremely significant compliment by reluctantly admitting at the 1989 press conference announcing the ban, that only 200 lives would be lost to asbestos by the year 2000, the same figure calculated by HEI, .0000048% of the population.

Nevertheless, despite all this accumulated scientific evidence and Reilly's admissions, legislation to extend existing federal regulations to 730,000 commercial and public buildings is still under active consideration in Congress. The legislation was filed by former U.S. Rep. James J. Florio (D. - N.J.), principal architect of asbestos legislation affecting schools and former chairman of the House Subcommittee on Transportation, Commerce, and Tourism, where both bills originated.

The bill was seconded by the Republican ranking member on the subcommittee, Rep. Norman Lent (R. - N.Y.). Both were co-sponsors of the Asbestos Hazard Emergency Response Act of 1986, which superseded an original 1983 EPA regulation. The regulation and AHERA cover an estimated 135,000 public and private schools at the elementary and secondary school level. Rep. Lent has declared:

"There is no doubt that. . .asbestos materials in our nation's schools is placing our children at considerable risk."

On the contrary, the accumulated scientific evidence leaves little or no doubt our children and occupants of the 20 percent of all public and commercial estimated to contain asbestos are at infinitely greater risk from EPA and Congress.

As Florio, recently elected governor of New Jersey, has admitted: "Cleanup is already underway in many commercial buildings. . . . This cleanup is proceeding with virtually no guidance. Without some guidance, this activity will become a health and environmental nightmare."

Moreover, as governor of New Jersey, he has already reduced expenditures for asbestos removal in schools enormously. That was in concurrence with a New Jersey state commission which basically agreed with the conclusions of the Royal Commission of Ontario back in 1984:

"Even a building whose air has a fiber level up to 10 times greater than that found in typical outdoor air would create a risk of fatality that is less than one-fiftieth of the risk of having a fatal automobile accident while driving to and from the building." Further, "[W]e deem the risk which asbestos poses to building occupants to be insignificant and therefore find that asbestos in building air will almost never pose a health risk to building occupants."

The New Jersey commission, in its own words, concluded:

"There are no documented cases of lung cancer associated with low-level asbestos exposure over a lifetime. . . . The estimated lung cancer mortality rates due to nonoccupational asbestos exposures were found to be about 10,000 times lower than the rates due to smoking."

But the asbestos cleanup continues, even without new legislation, and despite all the scientific evidence to the contrary. The cleanup is being driven, as the HEI report stated, in part by an "organized asbestos removal industry with a collective self-interest in removal" of $150 to $200 billion by the turn of the century.

Potential investors in the 20 or so asbestos abatement firms traded on the stock market are still being regaled with figures that show total revenues growing fivefold from $200 million-a-year in 1983 to $2.7 billion in 1987. Those figures could be multiplied several times over by the time the final HEI report comes out in 1992.

In the meantime, perception is, as always, in a close race with reality. Rifkin-Warnick, Inc., the leading asbestos business research firm, reported that asbestos abatement earnings in 1989 were $5.2 billion, almost twice the 1988 figures.

By further contrast, these earnings projected by Rifkin-Warnick for asbestos abatement almost directly parallel estimated budgets for the National Institutes of Health, the federal government's primary research organizations for cancer, heart disease, arthritis, and other diseases.

The figures provide a classic example of why linear thinking = insanity:

YEAR	ASBESTOS	NIH
1989	$5.2 B	$7.2 B
1990	$7.0 B	$7.6 B
1991	$8.3 B	$8.1 B
1992	$9.8 B	$8.7 B

Asbestos abatement revenues were anticipated in 1989-90 to really take off again in 1995, but may nose-dive as a consequence of EPA's apparent change in policy. But that's consistent with the up and down yo-yo thinking of an environmental movement willing to play roulette with human lives without consistent environmental thinking. And, in the case of asbestos that means the totally unnecessary deaths of workers doing an even more unnecessary job, ripping out asbestos rather than leaving it safely in place.

The comment of the late Senate Majority Leader Everett Dirksen is now sadly, indeed tragically, outdated: "A billion here, a billion there, after a while you get into real money." The amounts being tossed around now are in the trillions.

And the costs extend into the marketplace, making the necessities of building space either grossly overpriced—or undervalued. Florio himself has pointed out, "market forces are sharply devaluing properties known to contain asbestos"—an estimated 10 percent in Manhattan and even more in other cities.

For example, the Exxon Building in Manhattan was marked down $100 million and the price on the Arco building in Los Angeles was slashed $50 million when the buyers found they contained asbestos.

The total tab for asbestos abatement, counting abatement, depreciation, and lost wages of displaced building occupants, has been assessed at $2 trillion over the next three decades by Peter

MacDowell and Ehud Mouchey in the trade publication *Asbestos Issues* of June 1989.

And it's all due, despite Florio's protestations, to government prompting rather than private market forces. EPA is not entirely responsible for asbestos hysteria. It was prompted and encouraged by Congress and the environmental movement and their linear thinking. But EPA went beyond those demands with policy-making by public relations rather than science, the deliberate incitement of fear by press release.

3 Evidence

The United States government through EPA, has, in the words of Justice Oliver Wendell Holmes, been "crying fire in a crowded theater." The government's actions have been so irresponsible that, were it a private corporation, it could, and should be, arrested for incitement to riot.

And yet, despite all the evidence, as a *Wall Street Journal* editorial commented on September 18, 1989, "EPA hasn't made much effort to put the asbestos issue into proper scientific perspective in the wake of the Harvard report. Charles Elkins, an EPA spokesman, says that findings of the Harvard report is a position that 'we've taken for a good number of years.' This is surely news to a lot of private and public schools that have torn apart their budgets to satisfy the asbestos mandates."

Elkins' claim about EPA's position, to put it more bluntly than the *Journal*, is untrue, false, a lie.

So, too, are the words of another EPA spokesman quoted in *The Washington Times* December 11, 1989, as saying the agency believes "There is no evidence that any pesticide causes cancer."

That claim, too, is a lie.

They are not lies of any individual or spokesman, but the self-perpetuating lie of a government agency conceived in lies, nurtured in lies, and ripened in lies. Only Susan Sontag's stricture from *Illness as Metaphor* that the use of cancer as a metaphor has been so grossly abused it is therefore "morally impermissible to compare anything to cancer" spares EPA from the accusation of being the very thing it is presumably sparing the rest of us from: "EPA is the cancer of the United States."

31

Intellectual dishonesty, political cowardice, moral arrogance, and social corruption are evidently EPA's birthright and hallmark. All have combined to create an ecological nightmare which may be far worse than any predicted by the environmental movement.

The evidence is here in this book, evidence obtained in on-the-record interviews, from Freedom-of-Information requests, public documents, correspondence, quotes from individuals confirmed by at least two sources, and previously published records.

These are facts and they cannot be ignored.

There is no evidence anywhere that EPA has saved anyone from cancer. Indeed, especially in the cases of asbestos and the pesticide DDT, the evidence is conclusively to the contrary: EPA's bans of both are instead recklessly, wantonly, and criminally killing people for the sake of an environmental world view as outmoded as the Weltanschauung being tossed in the dustbin of history in the Communist nations.

For all the good EPA has done in preventing cancer, its budget should be paid by the Tobacco Institute rather than the public. For EPA has been the institute's loyal servant in fact, if not in name, by diverting attention from the only real source of preventable cancer, tobacco smoking.

And it all started under William Ruckelshaus when he banned DDT in his first term as administrator; and it went on in his second term when he played a significant role in implementing the original 1983 asbestos-in-schools regulation. He did so with the clear understanding that the asbestos policy had been formulated for one reason and one reason only: to dump the problem in the laps of local school administrators.

Further, EPA deliberately instigated mass hysteria, with Ruckelshaus' concurrence, under an overall public relations policy designed to "get the mothers to form vigilante mobs and storm the school committees."

Fear for children's' safety has been ruthlessly exploited by

EPA propaganda to "stampede local school administrators into inappropriate actions, that, in some instances can actually increase the hazard presented by asbestos-bearing products," Dr. Robert N. Sawyer, the country's leading asbestos-control expert, has warned. His voice is only one among dozens raised by qualified scientists which were, until Reilly's speech, consistently ignored. The result has been human tragedy and political corruption.

The evidence has long been found in scientific journals and is becoming more and more evident to the public. Even newspapers, news magazines, and news programs are beginning to report the vast harm caused by misguided asbestos policy. A pattern emerges from these news stories that anyone with common sense can see. The pattern underscores one central point: whole scale incompetence within the asbestos control industry—and illuminates another: the total lack of any scientific rationale for asbestos removal.

"Giving the industry the benefit of the doubt, maybe 10 percent of asbestos contractors are qualified," Neil Wilson, president of the National Association of Asbestos Abatement Contractors, has been quoted as saying.

That point is illustrated every day by new scandals. A second point flows from the first: A generally accepted way of looking at the world, the environmental perspective, is dying and another emerging. Human understanding as well as science works as Albert Einstein observed, "not by the establishment of greater and greater truths but by the elimination of more and more theories that are false."

The environmental theory or, more accurately, myth, of cancer causation which assumed that most cases were caused by exposure to minute exposures to a few carcinogens—asbestos, DDT, nitrites, etc.—has lost all credibility in the scientific community. Natural carcinogens have now been found to abound in the environment, in natural food, in our own bodies.

If everything or almost everything causes cancer, then just a few select carcinogens, cannot be singled out for blame, just because they are industrially generated or a product of modern civilization and, thus, offensive to members of the Audubon Society, Friends of the Earth, Environmental Defense Fund, and so forth.

Asbestos is apparently no longer at the top of EPA's hit list, but pesticides, PCV, PCBs, radiation, radon, etc, have become priority targets for the environmental movement. Indeed, Ralph Nader has gone into the business of selling radon detection kits.

Linear thinking has not gone away, and won't as long as people need scapegoats. But there is a growing realization that Mother Nature may not only be meaner than we think, but meaner than any industrial corporation. However, Mother Nature can't be sued, and even if she could doesn't have any "deep pockets" out of which financial judgements can be claimed.

The example of asbestos has made it clear that all risks are "relative," as Reilly has admitted and can only be judged in relation to one another, not by an absolute framed in linear dimensions.

"[B]y switching from city water to what the EPA considers contaminated water in the Silicon Valley, I could actually lower my risk [of cancer] by 300 percent," Dr. Daniel E. Koshland, editor of *Science*, published by the American Association for the Advancement of Science, wrote in the April 17, 1987, issue. "[T]he potassium in my body, which contains a radioactive isotope, has 1,500 times the radiation level of that from the atmosphere within 20 miles of a nuclear plant. . . . Should we, I wondered, abandon Superfund and find a substitute for potassium in the body?

"Lester Lave (James Higgins Professor of Economics, Carnegie-Mellon University), informs us that hazards around the house are half as likely to cause injuries as motor vehicle accidents and that asbestos poses a small risk in most appropriately constructed buildings.

"From recent reports on television and in newspapers, my impression is that we are dying like flies from exposures to toxic chemicals, nuclear power stations, drunken drivers and incompetent physicians," Koshland continued, "[Presumably] if one [could] avoid such hazards, and has a little help from an artificial organ here and there, dying seems to be pointless. . . . I am talking about living forever."

That, of course, is impossible, as Koshland observed, even if one were to "give up walking up and down stairs, drinking alcohol, living in Denver or other high-altitude locations, and be willing to sit in a rocking chair with a lead roof over my head and be fed amino acids intravenously."

Unfortunately, such comments and others coming from the most knowledgeable scientists in the country are only beginning to impact the media, the politicians, and through them the public. We live in a culture of fear, and that fear is inspired most by cancer.

James T. Patterson has observed in *The Dread Disease: Cancer and Modern American Culture*: "In no other nation have cancerphobia and 'wars' against cancer been more pronounced than in the United States." One reason may be that the fact that an ultimate cure—or preventive program—for cancer has not and may never be found offends the American can-do spirit. Nevertheless, the fact still remains, as one researcher observed back in 1949, "[While] great progress has been made in scientific understanding and medical treatment [of cancer], the final answer remains—and may always remain—like looking in a dark room for the black cat that isn't there."

Asbestos is no more to be feared than black cats. Nevertheless, asbestos hysteria is still spreading from schools to office buildings to apartment houses, even to rental units in two-family homes into New York City. Once the cat is out of the bag, as the cliche has it, it's difficult, if not impossible, to stuff it back in. And, as an inevitable consequence, an enormous asbestos racket has been created—because people can make money out of it.

Cancer, which has been used as metaphor for social evils for decades, has become a bipartisan source of corruption that transcends the entire political spectrum. The devils' bargain struck among politicians, bureaucrats, environmentalists, and a few scientists must be understood and resisted. Otherwise both human lives and billions of dollars will be wasted in a futile crusade against asbestos fueled by fanaticism and greed.

The dimensions of the corruption problem emerge graphically from the following sample of news stories:

● In New York, three of the principal executives of the 42 biggest asbestos removal firms in the metropolitan area have been found guilty of bribing an EPA inspector with almost $300,000. U.S. Labor Department officials indicate the three will testify against the other executives indicted. One of the companies, Big Apple, Inc. has been barred from receiving federal grants, contracts, grants, loans or assistance. "This is the first discretionary listing action in the nation," an EPA press release stated, "against a removal and demolition company for violations of the Clean Air Act for asbestos removal and control."

● In San Francisco Bay, dozens of homeless people have been hired to rip out asbestos from banks, churches, offices and hotels. The jobs pay $9 to $10-an-hour, but are often poorly supervised with workers being provided respirators but no protective suits. "It's sickening, in the literal sense of the word, because some of the workers will sicken and die," Terry Messman-Rucker of the Oakland Union of the Homeless said, a refrain that is being repeated all over the country.

The General Assembly's Office of Research in California has estimated the total cost of ripping out asbestos from all buildings containing the material to be $20 billion.

● In a Denver, Colorado suburb, 100 stores in a shopping mall, closed down by public health officials, were only allowed to reopen after millions of dollars worth of clothes and fabrics were bagged and dumped in a sanitary landfill. That was done because the store owners could not prove the materials did not contain asbestos. Total cost was $17 million including removal, lost rent, and reimbursement.

● Also in Denver, a federal grand jury has indicted three officers of asbestos control firms for bribing the hazardous materials manager of the public school system. The bribe was in the form of a campaign contribution to the manager's wife, a candidate for a seat in the state legislature. The penalty could be 30 years in jail and a $300,000 fine.

● In Detroit, San Francisco, New York, Atlanta—all over the country—owners of buildings suspected of containing asbestos are seeing the value of their buildings reassessed sharply downward. Appraisers estimate, as a rule of thumb, that all buildings constructed between the mid-thirties and mid-seventies are assumed to contain asbestos and have automatically been discounted 15 percent to 25 percent.

● In California, only six of the 540 contractors doing asbestos removal work were found to be registered with the state's Occupational Safety and Health Administration (Cal-OSHA).

● In San Antonio, 166 homeless workers have sued a contractor who hired them to rip out asbestos without providing adequate protection.

● In Kansas City, a qualified inspector made sure an asbestos removal job in a public school was carried out according to the rules—during the day. But, at night, the contractor would sneak in a bigger crew to do most of the work without any supervision.

● In New Jersey, the president of an asbestos removal firm was gunned down by seven bullets—one in the back of his head.

● Inspectors without even a high school diploma, derided as by key members of Congress as "five day wonders," are the only legally authoritative sources of advice and information available to deal with asbestos problems in school buildings. Twenty thousand of these inspectors, with little or no previous engineering, architectural or scientific training, are expected to be turned out of three-to-five day EPA approved courses.

● An asbestos removal worker at a Veterans' Administration Hospital complained to his foreman that the $12-$15 per hour paid him was considerably less than a Federal $25 minimum. The foreman poked one of the worker's eyes out. The foreman and his bosses in an asbestos removal firm have pleaded guilty to charges likely to lead to several years imprisonment. The foreman was the cousin of the workman whose eye was poked out.

● "Law [has] created a motive to be a criminal," Dan Millstone, a lawyer for the New York City Sanitation Department has said, in response to massive illegal dumping of asbestos. But that hasn't stopped the New York City Council from considering even more stringent asbestos legislation.

Lawyers have also become personally implicated in the corruption that seems inextricably associated with asbestos removal projects. The Raymark Company, a former asbestos producer, counter-sued attorneys who had won multi-million dollar suits against the company. Raymark contended two lawyers, one of them a former president of the American Trial Lawyers Association, ran a fraudulent program called the National Tire Workers Litigation Project. Technicians hired by the project examined some 100 to 150 workers a day in vans equipped with X-ray machines.

Diagnoses obtained through the X-rays and a questionnaire were used to win a $16 million settlement from Raymark on behalf of 7,000 workers. The average payout to the workers would have been $1,523, assuming the lawyers took only a one-third fee and no expenses—an unlikely contingency. Raymark contends it was duped into settlement by false information.

"In this court's view, this is the shabbiest form of claims

practice," said Judge Patrick F. Kelly of the Federal District Court in Wichita, Kansas. His comment came in refusing a motion from the defending lawyers to dismiss the Raymark allegations on the grounds the company had not presented sufficient evidence to support its allegations.

These are not isolated instances, but all part of a pattern, a pattern of corruption and exploitation, abetted all too often by the media.

4 Middle Class Enthusiasms ▄▄▄▄▄▄▄

The title of a *People* "news" story read: "Discovering Their Home is Walled with Asbestos, a Florida Family Flees the Dangerous Dust."

In the worst traditions of yellow journalism, originally described as comparable to a screaming woman running down the street with her throat cut, *USA Today* has shrieked: "There's a killer loose in the corridors, gyms and boiler rooms of 40,000 schools across the U.S.A."

EPA's effort to get rid of asbestos is turning into a racket on a par with Prohibition bootlegging. The health benefits are, at best, 17,500 times less than a ban on cigarettes. The price is not only massive social and political corruption, but also far worse, a waste of human life. Cigarettes cause up to 350,000 deaths a year. All current uses of asbestos cause, at most, less than 20 cases per year by EPA's own estimate, and that figure is believed by the scientific community to be greatly exaggerated.

We are destroying the village once again in order to save the people. But this time it is our village and our people.

All over the United States, fear of low-level exposure to asbestos is creating decreased property values, confusion, and panic. But what is worst and most important, this needless fear is causing a tragic and totally unnecessary loss of life among asbestos removal workers due to careless removal procedures and accidents. Many more lives may be lost among people living near asbestos dumps, according to EPA itself.

At stake is a loss of credibility in the environmental protection movement on a scale comparable to the general repudiation of the poverty programs of the '60s. We will have learned, once again, that not just morally bad but practically ineffective means do not justify good ends. The problems will remain, but there will be no faith in solutions.

And the very existence of EPA may be threatened in the backlash.

"You can't solve poverty by throwing millions at it," Richard Nixon said almost 20 years after the American public had revolted against corruption, waste, and just plain stupidity in the "War on Poverty." Then Nixon turned around and declared a "War on Cancer," costing many more wasted billions.

This "war on cancer"—along with the environmental movement—have been spared the massive criticism directed at the war on poverty for several reasons:

1. The public, and that includes members of the media as well as Congress, is terrified of cancer, and knows little if anything about science.

2. Until now the only systematic attack on regulatory bans of suspected carcinogens has been based on cost/risk analysis. Many Americans find that distasteful, however, because it apparently places a dollar value on human life or injuries.

3. The debate over effective control of carcinogens has been carried out for the most part in Washington within the limits imposed by a simple media story line which assigns industry, environmental activists, regulatory officials, and Congress stereotyped roles within a moralistic framework which rarely if ever takes new scientific evidence into account.

It has been a classic "inside the Beltway" story, in which almost all the action and the players have been located within Washington. Local and state officials have played almost insignificant roles, and taxpayers have never understood they are paying enormous amounts for almost non-existent health benefits.

But all that is changing, changing utterly.

EPA's own efforts first to remove asbestos from 135,000 public and private schools, and more recently from 730,000 private apartment and office buildings as well as public libraries, city halls, and police headquarters, is inevitably creating an entirely new climate of public opinion.

What EPA has done is to put cost/benefit analysis on the agenda of every school board and private building owner in the country. "Sure, absolutely," former EPA Administrator William Ruckleshaus has admitted. "We try to present them with what the risk is, what can be done to correct it—force them to go and find out if they have any risk and then force them to make a judgment as to what you should do.

"Now the question of whether you panic them," he added, "or whether the whole thing is going cause more harm than good; frankly, that's there any way you do it. I don't care whether the federal government does it or the local government does it, I think you've got the same problem."

But EPA, despite Administrator Reilly's 1990 American Enterprise Institute speech, has never officially admitted that's what it has been doing. Nor has it conceded that the agency itself really doesn't have the capacity to deal effectively with asbestos. It has preferred to foster an illusion of omnicompetence.

However, EPA's incompetence is manifest in the sampling of stories in dozens of newspapers and news programs all over the country cited earlier. Corruption, social, political, and criminal, is proving what knowledgeable scientists have known for years: EPA particularly, but also other federal regulatory agencies waging "war on cancer," literally don't know what they're doing.

These agencies have stumbled successfully through crisis after crisis over the years, because the public, the media, and Congress have wanted to believe they did actually know what they're doing. The nation in the '70s and '80s wanted to rid itself of carcinogens in the air, water and ground, just as it wanted to flush Communists out from under the beds in the '40s and '50s. But wanting to do something and having the ability to do it—and whether it can, or even should be done—are entirely different matters.

The realization is sinking in as the scandals multiply that something is as wrong with EPA now as with the Federal Bureau of Investigation (FBI) and the Central Intelligence Agency (CIA) in the '70s. Those agencies threatened our individual political freedoms in order to presumably protect us. Now EPA and other environmental agencies are also threatening community and social freedoms to presumably protect us. There is something perverse about this, as one typical editorial commented on the way EPA, the Occupational Safety and Health Administration (OSHA), and the Consumer Product Safety Commission (CPSC) are creating rackets to protect us from unseen enemies who may not, in fact, exist.

"The way the EPA has chosen to attack the problem seems almost perverse," the *Baltimore Evening Sun* has said. "Rather than establish objective standards for determining what constitutes a hazard, the agency in effect has left the decision about when to undertake expensive asbestos removal projects to the very contractors who stand to profit from the work. . .encouraging fly-by-night contractors boondoggles that reap billions at taxpayers expense."

Such editorials and news stories are already generating enormous political pressure. They are based on hard specific facts and evidence gathered by grand juries, prosecuting attorneys, and legislative hearings. The cumulative effects signal a fundamental sea change in public opinion.

A new moral dimension has be added to the environmental debate. Can we sacrifice the lives of asbestos abatement workers, particularly Hispanic, Caribbean, South American, and our own uneducated and poor black laborers, to soothe the largely illusory fears of predominantly middle-class white collar workers frightened by Public Broadcasting System (PBS) programs narrated by actresses such as Meryl Streep?

Should the lives of people living near asbestos dumps be sacrificed for the profit of unscrupulous contractors, ambitious politicians, sensation-seeking reporters, greedy lawyers, and power-hungry union leaders?

These questions did not long elude some members of Congress. For example, only a few months after EPA promulgated its regulations under AHERA in late 1987, signs of Congressional dissatisfaction with the agency's performance became apparent. Rep. Thomas A. Luken (D. - Ohio), chairman of the Subcommittee on Transportation, Tourism and Hazardous Waste, where AHERA originated, first used the derisive phrase "five day wonders" to describe the inspectors being turned out by EPA-approved courses. By the early summer of 1988 almost all the members of the subcommittee were supporting amendments to the AHERA.

Why would Congress think of reversing itself within months of the unanimous passage of an apparently universally accepted law?

The answer is as simple and dismaying as human nature. People wanted to do good for others but ended up doing well for themselves. A devil's bargain had been struck among environmentalists, advocacy lawyers, scientists, news reporters, and members of Congress. A fictitious epidemic of "asbestos poisoning" was created by the partners in the devils bargain. Originally they may well have wanted to help others. But environmental groups received generous contributions from new recruits, advocacy lawyers received generous fees, scientists received generous study grants, news reporters received generous accolades, and members of Congress received generous campaign donations. Soon they all found themselves only protecting themselves by trying to persuade millions they were endangered by asbestos. The real epidemic was fear, spread by scientific ignorance, bureaucratic bungling, political posturing, greedy lawyers, sensation-mongering reporters, and contractors chasing the almighty buck.

The facts began to rear their ugly heads. A few experts began to point out that the amount of asbestos present in the ambient air of most of the estimated 730,000 office buildings, apartment houses, and factories where the mineral was used from the 1930s through the mid-70s is negligible. In almost every instance, experts began to point out, the levels are comparable to or less than the amount in the air outside those buildings. Building owners who had been stampeded into removing asbestos began to realize that all the available scientific evidence made it abundantly clear such activity multiplied exposure levels many times over the presumably "safe" level set by EPA and OSHA.

The Clean Air Division of EPA estimated as far back as 1985 that at least 80 cases of cancer are caused each year from ripped out asbestos leaking out of landfill sites. By contrast, removal itself accounted for only one-half a case, and another half-case was produced by the use of asbestos in such life-saving applications as brake linings for trucks, planes, and automobiles.

Suspicious victims of the asbestos racket began to examine the record. What they found was disquieting. Asbestos removal is causing "more adverse health effects, including malignancies, than it is preventing," according to Dr. Robert Sawyer, who is both a medical doctor and engineer.

"Currently, there exists a common misconception. . .that the discovery of asbestos-bearing construction products automatically indicates serious contamination and exposure," he testified before an EPA hearing in 1985.

He said:

"The surge in demand for removals greatly reduces the probability that [building owners] will obtain a competent contractor, experienced workers, knowledgeable architectural advice and a safe removal operation. There is a limited supply of such qualified personnel, and demand has already exceeded supply.

"The basic problem is that you're causing cancer in abatement workers by allowing the unnecessary removal," Sawyer concluded.

Innocent lives are being traded to advance Washington careers.

But retribution is inevitable. What Sawyer has called the "third wave" of asbestos litigation is coming as inexorably as the tides. The asbestos removal businessmen of today won't escape punishment, just as the asbestos manufacturers and suppliers of the 1930s through the '70s didn't. Those companies have paid out tens of millions to workers exposed in the "first wave" of litigation and to family members in the "second wave." Yet those businessmen of an early generation had much less scientific reason to know, while asbestos was and is a great life saver when properly handled, that it is also a deadly killer unless properly controlled. There cannot be any such excuse for the profiteers in asbestos today.

Lawsuits are already being filed for asbestos removal workers. Unfortunately, the judgments handed down on behalf of those asbestos removal workers won't be nearly as generous as those for employees of asbestos suppliers. Most of those earlier supplier companies were major corporations which could afford big settlements, even at the cost of filing Chapter 11 bankruptcy proceedings in court, as the Johns Manville Corporation did. The vast majority of asbestos removal companies, however, are undercapitalized "mom & pop" operations which will probably go belly up before legal papers can be filed. Fewer than twenty of the asbestos control firms in the country are publicly traded on stock exchanges.

And, of course, building owners and contractors will bear major liability for harm done to people living near asbestos dump sites.

What makes this devil's bargain even more obscene is the fact that those who are primarily responsible will escape any punishment, even when the facts are generally recognized. One or two members of Congress might be embarrassed, but not seriously threatened at the polls. Reporters and editors will go on to other sensations. The bureaucrats will hide behind the "sovereign immunity" clause which protects government agencies and their officials from suits on the archaic assumption that "the king can do no wrong." The U.S. Navy has successfully used the clause to escape any responsibility for the deaths indisputably linked with asbestos, those of World War II shipyard workers. The workers were enveloped in clouds of asbestos because safety standards were neither set nor observed under wartime production demands. And the lawyers, even without the opportunity to sue the government, will continue to make money.

But building owners and the American public will have wasted upwards of $200 billion to cure a public health problem scientists consider agree to be negligible, indeed, almost non-existent.

The risk from asbestos in buildings has been compared to "smoking half-a-cigarette in a lifetime" by Sir Richard Doll of Oxford University, the world famed epidemiologist who conclusively established the relationship between cigarette smoking and lung cancer. He also published in the early 1950s the first studies linking asbestos with lung cancer.

Yet EPA is still telling both public and private school officials to spend at least $3.5 billion for asbestos removal and control, despite Administrator Reilly's posturing. Owners of other public buildings, schools, hospitals, libraries, and military installations will have lay out $61 billion according to a report submitted to Congress by EPA. That's a total of $64.5 billion. Asbestos control industry officials put the true cost at almost double those estimates, an actual $150 to $200 billion.

By contrast, $50 billion at the most would be needed to provide medical insurance to 22 million working Americans without coverage. So say critics of a bill to provide such coverage. Sen. Edward M. Kennedy (D. - Mass.), sponsor of the bill, places the cost at $29 billion.

Common sense gags at such comparisons—spending up to $200 billion to protect people from dangers as remote as being struck by lightning while arguing whether 22 million Americans subject to all the illnesses human flesh is heir to should be provided basic medical coverage. Our sense of proportion has been lost. Sen. Kennedy's plan has been compared to a Rolls Royce ambulance because the cost to hospitals for providing free service is less than $13 billion. Yet, we are using the equivalent of "Star Wars" systems to protect us from asbestos.

The arguments against unnecessary asbestos removal stand on the scientific merits alone. But economic considerations cannot be ignored. Traditional cost-benefit analysis is repugnant to many. But other comparisons have no such moral stigma, and, indeed, raise serious ethical questions for society: When does concern become obsessive waste? Are some lives more valuable than others?

For example, in considering how much asbestos control could or should cost society, these figures bear consideration:

● The bailout of the S&Ls is expected to cost as much as asbestos abatement, $500 billion.

● Total deposits for the seven largest banks in the country, The Bank of America, Citibank, Chase-Manhattan, Manufacturers-Hanover, Morgan Guaranty, and First Boston, were $282.2 billion. just 20 years ago in 1979.

● Wages of all employees, public and private, amounted to $248 billion in 1968, only thirty years ago.

● Corporate profits for all domestic American industries in 1986 were $220 billion.

● The total value of all commodities, food, machinery, scientific instruments, etc., sold overseas in 1986 was $217.3 billion.

● Two hundred billion dollars was enough to run the entire federal government for a year in the early '70s. That amount now represents one fifth of the federal budget for fiscal year 1991.

● Total farm income, including $11.8 billion in federal government subsidies, was $147 billion in 1986.

● The American balance of international payments was $141 billion in 1986.

● Total foreign assistance from 1950 to 1986—36 years in all—was $119 billion.

● Total corporate income taxes for 1988 were estimated at $117 billion.

Consider these figures on the lower end of the scale by contrast to asbestos control: They run from $3.5 billion for schools to $61 billion for private and public buildings other than schools, a total of $64.5 billion as estimated by EPA.

By comparison:

● Total wages of all government employees, federal, state and local, was $28.9 billion in 1985.

● Almost one half of the nation's bridges, 243,646 out of 575,607—all essential for trade, commerce, and private travel—are either structurally deficient or functionally obsolete. The Department of Transportation estimates $51 billion is needed to restore them to safe working order.

● Only three of the fifty largest corporations in the United States have assets in excess of $60 billion: General Motors, Ford and Exxon. Thirty-five of the fifty have assets of less than $20 billion.

● The total outstanding indebtedness of college and university students is, by coincidence, exactly $61 billion, the amount EPA estimates is needed to control asbestos in 500,000 private and public buildings other than schools.

And that figure doesn't take into the enormous loss in property values for buildings suspected of containing asbestos, an amount appraisers place at 15 to 25 percent of current value.

By contrast, in terms of public values, 90 to 95 percent of all college loans will be eventually repaid, with interest and a dividend—hundreds of thousands of better educated workers who can be expected to earn more and pay more in taxes.

By any rational economic standard, expenditures of this magnitude for asbestos control and removal make no sense. But, environmentalists and EPA protest this kind of cost-benefit analysis cannot be accepted because a dollar value cannot be assigned to human life.

But it is surely better morally as well as economically to save 10,000 lives than 1,000, and 100 than 10 lives. There has to be some balance struck between resources and results, otherwise resources will soon be exhausted and results negligible.

For example, used effectively, a fraction of the estimated $150 to $200 billion estimated for asbestos control could save and extend the lives of millions in Third World countries denied what we consider the most fundamental needs of food, housing and productive work.

The EPA ban on asbestos announced in July, 1989, has been denounced by scientists in Canada, Great Britain, Sweden, the United States and international organizations. For example, George Kleish, head of the Occupational Safety and Health Branch of the International Labor Organization (ILO), has said: "We have to live with asbestos. The ILO does not aim at prohibiting work with asbestos. We will have to concentrate on ways of reducing its health hazards."

Of particular concern is any effort to prohibit the use of asbestos in cement pipe, as already has been effectively done in several countries in Africa, Latin America, and Asia by the Agency for International Development (AID). The agency has refused to finance any manufacture of asbestos/cement pipe. Used as a strengthening agent, asbestos makes cement pipe a far cheaper alternative to pipe of cast iron or polyvinyl chloride—another suspected carcinogen at high doses. Such pipes would not only prevent tens of thousands of deaths from water-borne disease such as cholera and dysentery, but could irrigate fields raising food with water drawn from marshes which produce only malaria spreading mosquitos.

The very real health value of asbestos/cement pipe—over 200,000 miles of it are in use in the United States—is undeniable, and any risk of contracting cancer as a consequence so infinitesimal as to be absurd by any reasonable calculation. As J.J. Waddington, director of Environmental Health Services for the World Health Organization (WHO) has said: "The general [scientific] consensus is that imbibed asbestos via drinking water supplies no assessable risk to the health of the consumer."

By searing contrast, only 46% of the poorest fifth of the world's population enjoys clean water, according to the World Health Organization (WHO)—and are denied longer, safer lives as a result.

The very real life and death consequences for the poor of the Third World measured against an excessive concern over the almost insignificant health risks from asbestos in the United States should give any reasonable and morally responsible person pause for consideration. Even EPA officials, such as Fitzhugh Green, EPA's associate director for international affairs, recognize such a total ban raises "troubling moral issues."

Those issues aren't limited to the outside world, but are of paramount concern here at home where realistic consideration should be given, if not to cost/benefit analysis, then to relative risk assessment, an attempt to measure potential dangers and take effective action against them, a concept EPA Administrator Reilly has presumably adopted. And that's where a consideration of the health effects of cigarette smoking should be taken into account, both directly in connection with asbestos and in overall public health terms.

The synergistic relationship between cigarette smoking and asbestos has become universally recognized in the scientific/medical community over the years. Even the studies EPA has relied upon most heavily in its efforts to ban asbestos, done by Irving Selikoff, now retired from Mt. Sinai Hospital in New York, made it abundantly clear a generation ago that smoking increases the risk of contracting lung disease, especially cancer, 80-to-90 times over. The failure to bear that formula in mind over the years has made it impossible to give reasonable consideration of the health dangers associated with asbestos over the past 30 years.

And Selikoff has been constantly correcting himself in comments to the media over the years. But he has not been anxious to explain these constant revisions in the scientific and medical literature to his peers. That may be perhaps because questions have been raised and facts established that make his professional qualifications a matter of dispute, as we shall discuss in detail below.

In the meantime, during the late 1970s, Selikoff estimated 40,000 excess deaths a year would be caused by asbestos over the next decade. Over the next ten years, however, he kept reducing that estimate, first in half to 20,000, then to 12,000 and finally, in 1983, to 8,200.

And a 1984 report further illustrated the overwhelming

significance of cigarette smoking linked to asbestos exposure. The lung cancer death rates among asbestos workers who smoked was substantially higher than among nonsmoking asbestos workers.

Those figures were developed by Dr. Malcolm Ross of the U.S. Geological Survey for an article published in the *Journal of the American Society for Testing and Materials*. After reviewing the 36 published medical studies done all over the world and reviewing the mortality charts reported in the official *U.S. Vital Statistics*, Ross estimated "The average yearly asbestos mortality in the period 1967-77 was 522 deaths."

Selikoff's projection was 40,000-a-year for each of those 10 years.

That period, 1967-77, was the point at which asbestos-related disease could be expected to peak. The potential victims Selikoff studied were primarily World War II shipyard workers who had been exposed literally to clouds of the material while building warships and transports for the Navy. It wasn't until years later that serious efforts were made to set and enforce exposure limits to the material.

Since many of the workers were then in their late twenties or thirties in 1940-45, they would have been in their late 50s to early 70s during the 1967-1977 period. That's when the excess mortality rates projected by Selikoff on the basis of research done in the '50s and early '60s should have manifested themselves.

But, in fact, they did not.

The total actual excess mortality attributable to asbestos during the entire decade was 5,200, according to Ross's undisputed figures. That's an infinitesimal fraction of the aggregate 400,000 for the decade originally projected by Selikoff.

Ross's figure, 5,200, is approximately 80 times less than Selikoff's highest estimate.

Selikoff's original projections were based on studies done on workers exposed to totally unregulated exposures to asbestos during the '40s and early '50s. Most of the men were also smokers of unfiltered cigarettes. The Surgeon General, of course, concluded in the 1960s that even filtered cigarettes cause lung disease and cancer. However, common sense, expressed in such phrases as "coffin nails" to describe cigarettes, had recognized that fact, even in the 'forties. Smoking even then was an obvious danger, unlike asbestos.

But political agendas, especially at EPA, took precedence over common sense and scientific fact.

The real danger was the cigarettes, not the asbestos. It was as if, looking at the statistics linking highway deaths with alcohol abuse, society decided to outlaw automobiles rather than drunk driving. The cart was not only before the horse, but stupidity was also in the saddle.

But, to borrow a not completely inappropriate analogy from Communist thought, the Party line had to be set and the Party line had to be followed.

Political agendas all too often take precedence over common sense, especially when lawyers are involved. And the lawyers found they could successfully sue asbestos (but not tobacco) manufacturers. Nevertheless, the scientific facts are there and as irrefutable as any are at any time.

Selikoff himself has now said the risk from exposure from asbestos in buildings is extremely low. And British Ministry of Health study which concluded the cancer risk associated with working in a building containing some asbestos is 90 times lower than the cancer risk of a nonsmoker living with a smoker.

"If you take averages, there is very little risk," Selikoff himself is quoted as saying in the Fall, 1986 issue of *Environmental Action*, the magazine of a major environmental lobby.

But, of course, EPA did not take that observation into account in its ban on asbestos despite the fact that Selikoff's studies have been the agency's chief source of scientific justification for its actions. The reasons go beyond the usual bureaucratic intransigence. EPA is unique among federal agencies in not having an overall Congressional mandate.

The agency was not created by an act of Congress, but rather by an executive order of Richard Nixon, Reorganization Order Number Three of 1969. Subsequently, Congress passed almost a dozen different laws for the agency to administer. In consequence, the agency has almost 82 congressional committees and subcommittees to report to, all with different political agendas. The laws, written at different times and under differing conditions, are often at odds with one another. The Toxic Substances Act, for example, under which AHERA is administered, does not contain a definition of "toxicity," the level at which a substance becomes dangerous.

By failing to include a definition, Congress implicitly accepted the dogma of the environmental movement that there can be no safe level of exposure to a carcinogen. In the case of asbestos that translates into what has been called the "one fiber can kill" theory. That is the principal reason the Toxic Substances Division has successfully pressed for a total ban on all uses of asbestos in the future, announced in July, 1989.

But the record also shows EPA did modify the one-fiber-can-kill theory before the ban was announced. A Congressional committee, the House Appropriations Subcommittee on Housing and Urban Development (HUD) and Independent Agencies, which holds the agency's purse strings, forced the modification. It was pushed through, however, so quietly that almost no one noticed. The subcommittee, chaired by Rep. Edward Boland, forced the agency in 1986 to revise a guidance document for school officials to place more emphasis on containment in place rather than removal.

"There is no conclusive evidence of measurable adverse health effects from the inhalation of 'white asbestos'—the type found in 95 percent of buildings—at low nonoccupational exposure or from ingestion even at high concentrations," Boland wrote back in 1985.

"Recently the asbestos-in-schools program has received wide public attention," he added. "However, EPA's handling of the problem has not encouraged rational or responsible risk management. EPA technical guidance is lacking on how school officials are to evaluate risks and select appropriate corrective actions.

"The potential for risk and overreaction is tremendous—leading to wasted resources and possible higher asbestos concentrations when removal is undertaken where modest containment measures are appropriate. The committee urges EPA to provide more balanced and less emotional responses to asbestos-in-schools."

That message has been reiterated over and over again by responsible officials, but has been drowned out repeatedly by publicity-seeking officials and their sensation-seeking allies in the media.

EPA did, in fact, take some modest steps in the direction of rational responses to asbestos problems, primarily in cooperating with the Congressionally-chartered National Institute of Buildings Sciences (NIBS) in the development of model specifications for dealing with asbestos. The NIBS specifications, originally prepared in 1986 and re-issued in a revised form in the fall of 1988, were developed in cooperation with an 85-member task force, which included representatives of EPA as well as professionals from the nation's major building construction and repair associations.

"Eighty percent of all asbestos abatement projects are best handled by maintenance and repair rather than removal," David Harris, president of NIBS, said at the time. "There's no reason why maintenance and repair can't safely postpone removal until buildings are altered, renovated or demolished, when removal costs become merely part of a larger project."

But so-called "rip and skip" projects are still proliferating. One result is "a sense of frustration bordering on despair" for experts such as Dr. Morton Corn, head of the public hygiene division of Johns Hopkins University, a former assistant secretary of Labor for OSHA, and a co-author of the crucial *Science* quoted above.

The fundamental irony was removal efforts were and still are literally making the situation worse.

Thousands of air samples, reported from all over the world, have conclusively proved the highest peak levels of asbestos in building air were reached after asbestos removal projects were completed, Corn told an EPA hearing on August 7, 1987. Moreover, the ambient levels present in the buildings before removal was undertaken did not get back down to the same level until months, even years, after the asbestos was cleared safely according to EPA standards.

A Catch-22 situation—damned if you don't, and doubly damned if you do—was created.

"Two hypotheses are possible," Corn concluded. "The EPA has a vivid imagination which persists, despite evidence to the contrary, or the EPA has knowledge that such conditions do not occur and simply refuses to acknowledge the facts. I conclude the latter is valid; the agency has initiated a near-panic response to asbestos in non-occupational settings and will not acknowledge its overestimate of the magnitude of the risk."

EPA is, nevertheless, still, despite Reilly's 1990 protestations, not only trying to remove asbestos from buildings but also from all other commercial applications, including brake linings. It argued back in 1985, in announcing a proposed ban on all future uses of the mineral, that ban could save between 1,500 and 1,800 lives over the next 14 years, 107 to 128 per year.

By contrast, the Consumer Product Safety Commission (CPSC) has calculated 416 lives were lost in backyard swimming pool accidents in 1986, half of those children under the age of five. Yet the CPSC has never suggested backyard swimming pools be banned, despite the fact they cause as much as four times as many deaths as asbestos. Ironically, a ban CPSC did impose on asbestos in paints and putties sold to the general public led directly to the crash of the Challenger shuttle, as detailed in a later chapter.

Place these figures on the relative values of human lives in linear perspective:

A study done by the United States Administrative Conference, a government think tank, has estimated previous EPA bans of suspected carcinogens have cost as much as $8.5 million per life saved. The average fell in a range between $1.6 and $1.8 million per life saved.

A total ban on asbestos would impose costs on American society far higher than the EPA estimate of $65 billion to control asbestos in buildings, and easily as high if not higher than the industry calculation of $200 billion. But, in judging the value of a total asbestos ban, for purposes of comparison, assume the EPA estimate of 1,500 to 1,800 lives saved by the dawn of the 21st century is correct, although agency experts have admitted the calculations might be as many as 800 times too high. Assume further the actual complete cost of banning asbestos to be $65 to $200 billion, given that the real figure would probably be several multiples of that.

That is a maximum of 125 lives saved per year and a minimum of 107.

That brings the cost per life saved to a range of $36 to $43 million each at the $65 billion level and $110 to $130 million at the $200 billion level.

When the final ban was announced in July, 1989, the number of deaths avoided as a consequence had been slashed from 1,500 to 1,800 total to 200, one-seventh to one-eighth the original figure.

At that rates, the value of lives saved soars to $345 million at the $65 billion level and almost $1 billion at the $200 billion level

That is, in hard numbers, between $345,000,000 and $1,000,000,000 per life.

These figures, of course, are hypothetical, based entirely on extrapolations from the death rates expected of asbestos workers exposed to extremely high levels.

By contrast, the number of lives lost to tobacco, alcohol, firearms, and drugs are very real and substantiated. This is hard data, uncontested fact, as are these comparative death rate figures drawn from Public Health Service records:

Tobacco	320,000	Heroin	4,000
Alcohol	125,000	Cocaine	2,000
Auto Accidents	50,000	Asbestos (EPA maximum)	126
Killed by firearms	20,000	Asbestos (EPA minimum)	7
	Falls in the home	11,000	

Our concerns seem to be in direct inverse ratio to reality, but entirely consistent with linear thinking.

Cocaine alone kills anywhere from 20 to 100 percent more people in a single year than asbestos could in the worst case scenario over 14 years. More people die each and every year from opiates, including heroin, than asbestos could possibly kill by the year 2018.

The annual revenues of the tobacco industry, which is responsible for the lung cancer deaths of 350,000 Americans each year, were $37.7 billion in 1986—with the help of federal subsidies. On the other hand, EPA wants building owners, public and private, to spend $61 to $200 billion, two to six times that amount to save the lives of an unspecified number of building occupants—perhaps 50 a year.

The lives of those 50-a-year killed each are presumably worth as much as $130,000,000 million each. The lives of cigarette smokers are worth $11,781 by comparison, if you divide the 350,000 cases into the tobacco industry's $37.7 million in royalties.

This goes far beyond cost/analysis into the realm of what Corn has called the "surrealistic." The value of human life itself becomes meaningless when vast amounts of resources are marshalled to save a hundred or so lives and little or nothing to save hundreds of thousands. The very concept of equality before the law becomes a farce.

EPA itself has estimated that 4,414 cancer cases per year would actually be caused by careless asbestos removal jobs, since 75 percent of all asbestos removal jobs are sloppily performed. By contrast, maintenance programs would prevent a net of 104 cases. Are the lives of asbestos removal workers 400 times less valuable than those of maintenance workers? Is that because many asbestos removal workers are Third World immigrants, and most maintenance workers are Americans born of European stock?

In this surrealistic realm, decisions are made on the basis of what is *not* known, rather than what is known. Arguments such as this exchange between William Tucker, a writer, and William Butler, attorney for *Harper's* Magazine, recorded in 1979 illustrate the absurdity:

Butler: "You can't prove that something doesn't cause cancer, because it may cause cancer, only no one has proved it yet."

Tucker: "But by that reasoning, you can't prove that dragons don't exist."

Butler: "That's right. You can't prove that dragons don't exist."

This is the realm of madness—and linear thinking—I am trying to address. In the words of an editorial in *The Detroit News* commenting on a news story I had written:

"Reporter Bennett made some telling additional points that go to the heart of the whole environmental debate in this country. Environmental extremism is difficult to defend against because scientists can't prove a negative—they can't prove that cancer or other risks from any given substance *don't* exist. So irresponsible extremists can always play on the fears of risks that *might* exist.

"The asbestos case also rests on the 'one fiber can kill' argument. That is, no amount of asbestos can be allowed in the environment, because as little as one fiber can supposedly make the body go haywire. That can't be proved or disproved, but for extremists seeking to impose their values on society it has the virtue of forestalling attempts to weigh risks versus benefits. As Mr. Bennett pointed out, environmental activism exists **not** to be satisfied, at least on environmental concerns. Many who profess environmental concerns actually are pursuing larger political goals—to remake society along the lines they prefer. Asbestos is a classic example of that strategy. In that it uses parental hysteria over the safety of children, it is all the more reprehensible."

But, despite all these arguments, asbestos must be banned, EPA still contends.

Why?

Dr. John A. Moore, assistant administrator of EPA for toxic substances, in testimony before Congress, has supplied perhaps the most candid—if least logical—reason: "If we can't ban asbestos, we can't ban anything."

For that's what EPA does: ban, prohibit, interdict.

EPA does not regulate, which could be defined as "to govern by rule or established mode, to govern by or subject to certain rules and restrictions; to direct; to put and keep in good order; to control and act to cause positively."

Therefore, one point should be clear: EPA bans, it does not regulate. And it bans usually over the protests of the scientists most knowledgeable about uses and dangers associated with chemicals.

Dr. Bruce Ames, chairman of the department of biochemistry at the University of California at Berkeley, was one of the first proponents of reducing the limits of workers' exposure to ethylene dibromide (EDB), a soil fumigant banned by EPA in 1984 because it caused cancer in rats. But Ames urged the agency not to ban EDB unless it recommended an equally effective substitute to help farmers fight crop pests—the kind of risk-benefit question also raised about asbestos.

EPA went ahead despite the fact that the only equally effective method is irradiation, a highly controversial method of exposing fruits and vegetables to low levels of radiation.

Ames is also the developer of the Ames test, a simple, inexpensive test for identifying mutagens (gene-changing agents). The test has, for all practical purposes, replaced the notoriously unreliable animal tests, principally in rats and guinea pigs, for identifying carcinogens, which are usually mutagens. The Ames test costs about $500; animal tests an average of $1 million. The results of both are considered by researchers as equally good indicators.

In creating the test, Ames totally upset the premise on which the environmental movement rests, that carcinogens are few in number, the product of modern industrial civilization, and easily replaceable.

"Far from being a rarity," Ames, an early proponent of the environmental movement, has said, carcinogens are "literally everywhere." Carcinogens abound naturally in such foods as uncooked corn, nuts, beets, celery, lettuce, spinach, radishes, mushrooms and rhubarb.

"Cooking our food also generates mutagens and carcinogens, as all browned and burned materials contains them," he has written in an article for *Science*.

And "it takes a Los Angeles resident a year of breathing smog to equal the burned material that a heavy smoker inhales in one day." Ames' article continued, "And we eat even more browned and burned material than the heavy smoker inhales, though we don't know the risk from this."

Carcinogenesis may, in fact, be caused, in a sense, by life itself. The only direct cause-and-effect relationship between exposure to a carcinogen and the development of cancer that has been clearly established is smoking, particularly cigarettes. Cigarette smoking, in conjunction with high level exposure to other carcinogens such as asbestos can multiply the chances of developing cancer many times over. But not all smokers develop cancer, and all of us are exposed every day to literally thousands of carcinogens in our food, as well as in the environment. High levels of exposure to these carcinogens unquestionably increase risk, but low levels seem paradoxically necessary for the maintenance of health.

Arsenic, for example, is the deadly "poison of the Borgias" that Rachel Carson compared the pesticide DDT to in her epochal book, *Silent Spring*. Arsenic is, in fact, a carcinogen, but it also one of the 21 essential vitamins and minerals needed to maintain human health, one usually obtained by eating lobster, shrimp and other foods.

"People hear carcinogen and they get so excited," Ames has said, "But you have to think of amounts. Sunlight is a carcinogen, but you don't tell your friend not to walk across the street in the sun."

Other carcinogens can't be banned any more than sunlight. There are no currently available substitutes for asbestos that are as effective, as has been tragically demonstrated by the CPSC ban on asbestos in paints and putties that led to the Challenger crash.

"When we banned certain products that contain asbestos," according to one federal insider, Barbara Franklin Hackman, a CPSC commissioner under Jimmy Carter, "the truth of the matter is we lacked information to evaluate the safety of the substitute materials."

That lack of knowledge led, with the inexorable inevitability of a Greek tragedy, to the crash of the Challenger shuttle and the death of its seven member crew.

The logical impossibility of proving a negative—and the invincible ignorance of politicians and bureaucrats— covered up the true horror of the crash, despite the best efforts of a Nobel Prize winning scientist.

It was an accident waiting to happen—a trajectory without an orbit—linear thinking without thought.

A little neglect may breed mischief. . .
For want of a nail the shoe was lost;
for want of a shoe the horse was lost;
for want of a horse the rider was lost,
and for want of a rider a kingdom was lost.
—Benjamin Franklin

5 Safety Kills: The Challenger Tragedy ■■■

An excessive concern over the possible danger of asbestos in personal hair dryers led directly to the crash of the space shuttle Challenger and the deaths of its seven crew members.

A 1977 ban on asbestos in hair dryers triggered a series of decisions that made unavailable the asbestos putty which had safely sealed the spaceship engines, and led directly to the crash.

The substitute, designed for jet planes rather than rocket engines, lacked the insulating powers of the original. The seals on the engine could not resist the abnormally cold weather they were subjected to for days in advance of the launch.

The Challenger crew was doomed.

They were martyrs to extremism that has taken the form of witch trials in the past and regulatory hysteria in our time.

The Challenger story is, in many ways, a classic Greek tragedy played out against the backdrop of outer space. But, as in most "tragedies" reported in the news, there are no heroes, only victims.

The heroes of great tragedy were great men brought low by a fatal flaw in their character. The tragedy of the Challenger was that seven living representatives of the American pioneer tradition were smashed to earth by political cowardice and legal arrogance that destroyed the spaceship Challenger in the name of "Spaceship Earth." The Challenger crashed because the Consumer Product Safety Commission (CPSC) had decided to ban asbestos in paints and putties sold the general public. The commission was reacting

65

primarily to TV news accounts about the presumed cancer danger associated with asbestos in the personal hair dryers used by the blow-dried generation. What the commission did not take into account was the far more real danger from fire and freezing that asbestos prevents. Those were only two of the dangers the putty was used to guard against in sealing the O-Rings on the Challenger, primarily preventing gases seeping through cracks or blowholes.

Nor did the commission question the assumption of Ralph Nader and others that laws and regulations could produce technological change.

Most environmental regulations contain a "technology-forcing component," as David Dickson has written in *The New Politics of Cancer*, "based on the idea that if a corporation was obliged to meet certain environmental and health standards, it would carry out research to develop new technology to meet those standards that it might not have done if technological choices had been left to the criteria of market demand and profitability alone.

"A typical example," observed Dickson, European correspondent for *Science* and former American correspondent for *The Times of London*, "is the catalytic converter for car exhausts, strongly opposed by Detroit when first proposed as beyond the scope of existing technology at a reasonable cost, later adopted with little difficulty when the auto industry found more efficient ways to meet government air pollution standards."

The Detroit example, however, is very far from typical. The auto industry knew what was being demanded of it, and fought long and hard for the time needed to develop the necessary new technology. The issues were fully and openly debated in Congress, the media, and the car-buying public. But no one understood at the time that the decision was made to ban asbestos in consumer products that it would lead directly to the Challenger crash.

The National Aeronautics and Space Administration (NASA) was forced, without knowing the reasons why, to use "the wrong stuff" in the Challenger. The result was a high frontier tragedy.

The facts are known but scattered in news accounts, articles, and books, never completely assembled as an indictment of the Washington system that produced this useless waste of life and threat to the space program protecting our national security.

The truth wasn't covered up so much as it was ignored. It was ignored because it demonstrated all too uncomfortably the inability of the system to anticipate consequences of narrow political decisions taken to appease clamoring interest groups. Ironically, it illustrates more vividly than any recent example cited by environmental groups the holistic interrelated nature of life. It is a tragic parody of the concept of "Spaceship Earth," a metaphor appropriated by those who arrogantly refuse to trust the trained knowledge of the scientists and engineers who made the space program possible.

"Safety Kills" is now a metaphor for the hubris, the inordinate pride, and exaggerated self-importance of the environmental movement.

This is the story of the events that led up to the accident that was waiting to happen.

The lives of tens, if not hundreds of thousands, have probably been saved by asbestos used as a fire retardant in buildings and ships and as a braking material on automobiles, trucks, and planes. Few, if any, would deny that common sense observation, despite the fact that logically a negative can't be proven. The material has demonstrated its ability to withstand flame and stress in too many thousands of laboratory tests to leave much doubt about its value.

Further evidence drawn from the records and reports of fire fighters, police officers, and pathologists shows that asbestos has saved infinitely more lives than it has cost. Taking the material out of a proven technological use, particularly engineering applications which must be reliable, is unjustifiably imprudent. For that reason, the American Association of Automobile Engineers has adamantly opposed attempts to ban asbestos in brake linings.

Their decision is based on the "principles of probability." Since all science is based on numbers, decisions are usually based on the balance of the numbers, one way or another. Merely theoretical arguments such as the impossibility of proving a negative or finally establishing the non-existence of dragons are ignored. Action is taken on prudent interpretation of what is actually known and established, facts and reason, rather than emotions such as fear, ambition, and power lust.

However, fear, ambition, and power lust covered up the origin of the Challenger crash because the Washington establishment—not just the top management of the NASA, but also the White House—really didn't want to look into it too deeply. Politics had created a climate in which the accident was inevitable.

The immediate cause of the accident was a "lack of communication between engineers and management," Richard P. Feynman, a Nobel Prize winning physicist and the top scientist on the Presidential panel investigating the crash, wrote in *Physics Today*, February, 1988.

The putty protecting the critical O-rings in the Challenger's two booster engines against the fiery rocket exhaust had been failing as far back as 1981 because of a lack of effective asbestos insulation. The laws of probability were saying that sooner or later there had to be an accident. But "NASA had no system for fixing the problem, even though engineers were writing letters like 'HELP!,' and 'This is a RED ALERT!'" Feynman concluded.

Production personnel at Morton-Thiokol, manufacturer of the engines, thought by putting on "management hats" they not only outranked "engineering hats," but presumably suspended the laws of probability, Feynman said.

Feynman came to these conclusions after doing what Sen. Ernest Hollings (D. - S.C.) said was the only thing that would produce answers about the crash.

"So who have ya got there on your commission?," Hollings challenged the chairman of the Presidential commission, William P. Rogers, at a Congressional hearing. "Ya got a couple of astronauts, a Nobel prize winner, a general, some businessmen, and a couple of lawyers. What you really need are gumshoes who will go right down there at Kennedy, eating lunch with the very guys who do the work on the shuttle."

That was exactly the kind of investigation Feynman, the Nobel prize winner, conducted. But he was dealing with politicians, not professional fact finders in reporting back his findings. And he was outflanked. Feynman submitted his findings, which highlighted the effect of cold—and the lack of proper insulation which had been earlier provided by asbestos—on the rubber in the O-Rings, to Al Keel, the executive officer of the commission. "He [Keel] told me on the telephone that he had received it and he would show it to everyone." But when the commissioners met to go over the final report with Rogers, the chairman and a former U.S. Secretary of State, the focus was placed on style rather than substance. Rogers did not permit discussion of the causes of the crash at commission meetings, nor "where to go next" scenarios drawn from the various perspectives of the astronauts, lawyers, industrialists, and engineers on the panel.

"In our meetings," Feynman wrote, "all we ever did was what they called 'wordsmithing'—correcting punctuation, refining phrases and so on. We never had a real discussion of ideas!

"Besides the wordsmithing, we discussed the typography and the color of the cover. At each meeting we were asked to vote [on the color]. . . . We voted red. It came out blue." Then, after one of the meetings Feynman mentioned to Sally Ride, the former astronaut, his special report and realized she hadn't read it.

"So I said to Mr. Keel, 'Sally hasn't seen my report.'

"He says to his secretary, 'Oh, make a copy of Mr. Feynman's report and give it to Ms. Ride.'"

Then Feynman discovered another commissioner hadn't seen the report.

"'Make a copy and give it to Mr. . . .'

"I finally caught on, so I said, 'Mr. Keel, I don't think anyone has seen my report.'

"So he said to his secretary, 'Make a copy for all the commissioners and give it to them.'

"Then I said, 'I thought you told me you [already] showed it to everybody.'

"'I meant I showed it to the entire staff.'"

"Needless to say," a now wary Feynman wrote later, "when I asked the members of the staff about it, none of them had seen it either."

But Feynman, although he was smart enough to detect the political web being woven, lacked the political skill or the instinct for the jugular to penetrate and expose it. That was prefigured during one of his earliest interviews with officials responsible for the crash. He had been examining a Morton-Thiokol executive who had changed his mind about the shuttle launch after being told to take off his "engineering" hat and put on his "management" hat.

"Don't you understand the principles of probability," Feynman asked the executive and then backed off because he had permitted his scientific approach to be compromised by political considerations.

"Suddenly, I had this feeling of the Inquisition," he wrote. Feynman had remembered what Rogers had told the commissioners at the outset of their investigation. "We ought to be careful about these people, whose careers depend on us." Rogers had said: "We have all the advantages. We're sitting up here; they're sitting down there. They have to answer our questions; we don't have to answer their questions. It isn't fair."

It might not be fair, but it was what the commission was presumably set up to do: find all the answers, identify all the mistaken judgments so no such accident could happen again—and, if need be, send those responsible for any criminal liability to jail.

Rogers' advice was that of a politician, not a prosecutor. If any member of the commission was a professional investigator, he or she would have quit right then.

Feynman did persist in having his findings published in the commission's report. But he compromised. "Although some of the members felt strongly that it ought to go in the main report," he wrote, "I thought I'd compromise and let it go in an appendix." Feynman was persuaded to have his report included as an appendix, rather than part of the main report, solely for technical reasons. "All we were doing was this wordsmithing stuff on what was already written–not adding anything new," he wrote, "It had to be put into the document computer system, which was quite elaborate, and very good, but different from the computer system I had written my report on back home." The lack of computer compatibility was apparently worth more than seven lives and the future of the space program.

Feynman was to temporize again. The commission made nine recommendations at its final meeting. The next day Feynman was "standing around Rogers' office when he [Rogers] says, 'I thought we would add a tenth recommendation: 'The commission strongly recommends that NASA continues to receive the support of the Administration and the nation.'"

Feynman protested vehemently: "It wasn't in our directive from the President. We were only to look at the accident, find out what caused it, and make recommendations to avoid such accidents in the future."

"What a mistake it looks like–just like one of those NASA reports," Feynman thought, "like the one I'd seen back in the beginning with the contradictory bullets. There's all these troubles, but in the end we recommend to keep on flying! I knew I didn't like it! Furthermore, we hadn't discussed it at a meeting!"

Feynman continued to protest, but was informed by Rogers he had been outvoted by all the other commissioners in a telephone poll.

Returning home, he got the idea to call some of the other commissioners: "I'll call them A, B and C.

"I call A: He says, 'What tenth recommendation?' I call B: He says, 'Tenth recommendation, what are you talking about?' I call C: He says, 'Don't you remember, you dope? I was in the office and I don't see anything wrong with it." Feynman continued to argue back and forth between Rogers and the commissioners. He was also angry about being "railroaded into modifying my report, even though it was going to appear only as an appendix."

He finally decided on a formal protest, and sent a telegram to Rogers, which read: "Please take my signature off the front page of the report unless two things occur: 1) There is no tenth recommenda-

tion, and 2) my report appears as a appendix without modification."
The arguments continued but none had "any effect" on Feynman, he
claimed: "The arguments were all kinds of crazy things. For example,
'If you don't accept the tenth recommendation, they're not going to
accept the compromise already made about putting your report in as
an appendix. I didn't worry about that one, because I didn't have to
sign the main report, and I could always put the report out by myself.

"Another argument was that they noticed I was always talking
to the press and they would claim I was doing this as a publicity stunt
to sell my book [presumably his report in book form]. That one made
me smile, because I could imagine the laughter that would produce
from my friends back home. I knew that nobody I cared about would
believe it."

Then, after giving all his reasons for not caving in, Feynman
wrote: "But finally, I did compromise. I said: Instead of making it a
[tenth] recommendation, just make it a concluding thought and
change the wording from 'strongly recommends' to simply 'urges.'"
They accepted that.

"A little bit later, Mr. Keel calls me up: 'Can we say strongly
urges?' "I said, 'No. Just urges.'" Word that Feynman had dissented
with Rogers leaked to the press just before the commission report
was officially released. Feynman held his own press conference "and
found myself repeating I don't have any problem with Mr. Rogers."
He said so, Feynman explained, because "I like him and that he's a
genuinely fine fellow...[although] he's such a good politician that he
knew how to make me like him. . . . As a scientist I don't know whether
my evidence is complete." In other words, Feynman didn't tell the
truth, the whole truth, and all the truth as witnesses before the
commission were required to do—and as scientists uphold as their
highest obligation.

The Nobel Prize winner was a prisoner of politics.

And so was NASA.

It took Feynman almost three years after the release of the
Challenger Commission report on June 6, 1986—a scant five months
after the actual crash on January 28, 1986—to realize just how much

he had been a prisoner of politics. He revealed how in a series of biographical essays, *What Do You Care What Other People Think?*, published soon after his death in 1989. And he learned you can't entirely trust even those you trust, because of their political vulnerability.

From the start, the member of the commission Feynman became closest to was Air Force General Donald J. Kuytna. As the first informal session of the commission broke up, Fenyman overheard Kuytna ask where the nearest Metro stop was.

"I thought: "This guy I'm going to get along with fine; he's dressed so fancy, but inside he's straight. He's not the kind of general who's looking for his driver and special car; he goes back to the Pentagon by Metro. Right away I liked him and over the course of the commission I found his judgment was excellent."

But Kuytna was still a Pentagon general, and he had to operate within the political limits of Washington. That meant he couldn't be completely straight with anyone.

Feynman had instinctively understood these political constraints when he was first asked to join the commission. "When I first heard the investigation would be in Washington, my immediate reaction was not to do it. I have a principle of not going anywhere near Washington or having anything to do with the government. So, my immediate reaction was: How do I get out of this?" His wife persuaded him he should go. "If you don't do it, there'll be 12 people—all in a group—going around from place to place together." But with Feynman in the group, 11 would going around from place to place together—"while the 12th one runs around all over the place, checking out all kinds of unusual things. There probably won't be anything but if there is, you'll find it. There just isn't anyone else who can do that like you can."

"Being very immodest, I believed her."

It turned out there was something, but Feynman got a hold only of a part of it, and then by indirection and incompletely: "Another thing I understand better now has to do with where the idea came from that cold affects the O-rings," he reminisced in *What Do*

You Care What Other People Think? "It was General Kuytna who called me up and said, 'I was working on my carburetor, and I was thinking: 'What is the effect of cold on the O-rings?'

"Well, it turns out one of NASA's own astronauts told him [Kutyna] there was information somewhere in the works of NASA, that the O-rings had no resilience whatever at low temperatures, and NASA wasn't saying anything about it.

"But General Kuytna had the career of that astronaut to worry about, so the *real* thing [Feynman's italics] he was thinking as he worked on the carburetor was, 'How can I get this information out without jeopardizing my astronaut friend?' His solution was to get the professor excited about it and it worked perfectly."

Feynman, even years later, still didn't fully understand what had happened. But he had learned a fundamental lesson: Never trust a politician or a political appointee. And he began to understand why he had greatly upset Keel with what was intended as an innocent question at the outset of the commission inquiry.

"I asked him [Keel] a particular question, which [in retrospect] could be considered a grave insult. The only way to have real success in science, the field I'm familiar with, is to describe the evidence very carefully without regard for the way you think it should be. If you have a theory, you must explain what's good and what's bad about it equally. In science, you learn a standard kind of integrity and honesty.

"When I see a congressman giving his opinion on something, I always wonder if it represents his *real* opinion [Feynman's italics] or if it represents an opinion he's designed in order to be elected. It seems to be a central problem for politicians. So, I often wonder, what is the relationship of integrity to working in the government?

"Now, Dr. Keel started out by telling me he had a degree in physics. I always assume that everyone in physics has integrity—perhaps I'm naive about that—so I must have asked him a question I often think about. 'How can a man of integrity get along in Washington?'

"It's very easy to read that question another way: 'Since

you're getting along in Washington, you can't be a man of integrity.'"

That was a conclusion drawn by many journalists reporting the Challenger story.

For example, Joseph Trento, who covered NASA for 15 years for Cable News Network and other news organizations, has confirmed in *Prescription for Disaster: From the Glory of Apollo to the Betrayal of the Challenger,* a lack of sufficient asbestos insulation was the primary technical cause of the crash.

But the technical failure occurred only because of politics, according to Trento. An independent Office of Quality Assurance had been established by J. Keith Glennan, NASA's first administrator, "to serve as an outside auditor, and go directly to the administrator's office to correct a problem." However, the independent auditor's office was systematically underfinanced and downgraded over the years.

"By the end of 1980," Trento reported, "NASA's capability to technically verify any contractor's work had all but vanished. Quality assurance had been so downgraded by 1985 that a NASA field center like Marshall [where the shuttle was put together] could keep a lid on a problem." That's just what happened with the Challenger. Neither the NASA astronauts nor the NASA top management had "any way of knowing that a 1977 Consumer Product Safety Commission (CPSC) ruling banning asbestos in certain paints and putties would have a tragic effect on the flight," as Trento reported.

"NASA had used an 'off-the-shelf' putty manufactured by the Fuller-O'Brien Company of San Francisco to help seal the field joints of the SRBs (Solid Rocket Fuel Boosters) through the first 10 missions," Trento wrote. "The Fuller-O'Brien Company, fearful of legal action because of the asbestos ban, stopped manufacturing the asbestos based putty.

"NASA began buying a different putty from a New Jersey company," Trento continued. "The SRB experts at Marshall noted the new putty did not seem to seal the joints as well as the old putty, but they continued to use it anyway." What Feynman was later to describe as "Russian roulette" and "a perpetual movement headed for trouble" had begun with the CPSC ban on asbestos in paints, putties and spackling compounds in 1977. The decision was prompted by news stories, fanned by consumer groups, that asbestos could be found in many products, including such innocuous items has hair dryers. Studies indicating asbestos could be carcinogenic were cited, but without reference to the fact that the presumed victims had been exposed to extremely high levels, and were usually cigarette smokers.

One member of the CPSC was troubled by the decision even back then. Barbara Hackman Franklin was uneasy over the "difficult balancing process of risk assessment." At a 1978 asbestos conference, Franklin observed government regulators are "charged with making decisions that are in the public interest. This means weighing the need for adequate consumer safety against the possibility of greatly increased costs and burdens for industry, and ultimately the consumer.

"The consequences of whatever we do must be considered, so that we do not cause unnecessary economic upheaval." Unfortunately, she apparently didn't understand that the CSPC's actions could—and inevitably would—endanger lives as well. CPSC, the environmental movement, and the media were crying fire in a crowded theater, ironically and tragically stripped of the one material that could prevent the fire from spreading long enough for the patrons to get out.

Hackman's voice went unheard before the Challenger accident. Just as unfortunately, the management at NASA and Morton-Thikol apparently never thought to question why they had been deprived of a highly successful technology that had worked for almost 15 years.

Warnings that heedless bans of asbestos could result in the loss of human life had been sounded well before the crash. It was only a matter of time before a prophecy made by Dr. Malcolm Ross of the U.S. Geological Survey in a seminal survey of asbestos-related disease published in 1984 came true: "Possible problems can occur with substitutes, for example, with the replacement of chrysotile [asbestos] in drum brake linings. The chance of increased automobile accidents due to a possible inferior substitute must be weighed against the probability of anyone being harmed by the small amounts of chrysotile emitted by drum brakes." Other warnings were being sounded by NASA engineers—and also ignored. As early as 1981, the NASA engineers were becoming worried about erosion marks seen on examination of the O-ring seals after the re-usable shuttle returned from space missions.

On July 17, 1985, Irving Davids, an engineer in the solid fuel rocket division of NASA, sent a memo to his direct line superiors—and Jess Moore, the agency's associate administrator for space flight, which read, in part: "The primary suspect as for the cause for the erosion on the primary O-ring seals is the putty used. . . . It is important to note that after the STS-10 [the tenth shuttle launch], the manufacturers of the putty went out of business and a new putty manufacturer was contracted. The new putty is believed to be more susceptible to environmental effects such as moisture which makes the putty more tacky.

"There are varying options being considered such as removal of the asbestos, varying the putty configurations to prevent the jetting effect, use of a putty made by a Canadian manufacturer, which includes asbestos and various combinations of putty and grease.

"Thiokol is seriously considering the deletion of the putty. . . since they believe the putty is the prime cause of the erosion."

Other such memos were circulating in NASA. One went to the head of the budget review section on July 23, 1985 from Richard Cook, a budget analyst who wrote: "Engineers have not yet determined the cause of the problem. Candidates include the use of a new type of putty (the putty formerly used was removed from the market by NASA because it contained asbestos)."

The authors of the memos were obviously confused or misinformed about who had banned the original putty—but they were on the right track. What actually happened, following the CPSC ban, was that the manufacturer of the original sealant putty, the Fuller-O'Brien Company of San Francisco, decided to stop production since the formula contained asbestos. For, of course, the environmental lobby then and now was touting the "one fiber can kill" theory of cancer causation.

"We didn't want any law suits," explained Tim Kelly, Fuller O'Brien's vice president for technology. "We wanted asbestos off the premises"—and that included all the company's product lines. The likelihood of any respirable asbestos escaping from the putty would, as a matter of reason, seem extremely remote. But reason was obviously not prevailing, and since the putty represented only one percent of the company's product line, it was obviously expendable.

The CPSC ban did not apply to industrial purchasers, such as Morton Thiokol, nor government buyers, such as NASA. But Fuller-O'Brien anticipated that lawsuits—and a total ban on asbestos—would be forthcoming, and just such a total ban was proposed by EPA about the time the Challenger and its crew plunged into the Atlantic ocean.

So, Fuller-O'Brien, back in 1978, decided to get asbestos out of all its products in one swoop. The company was able to find substitutes for asbestos in all its products—with the sole exception of the putty used, not only in the Challenger shuttle, but also in the Titan 34D rocket, made by Martin Marietta, used to launch spy satellites into orbit.

There simply was no substitute for the asbestos in the putty. However, since the share of business commanded by the material was only one percent of Fuller-O'Brien revenues, it was jettisoned.

The company did make a sufficiently large backlog of the putty to satisfy old customers such as Morton-Thiokol up until 1983-84 before shutting down production. Then the formula—dubbed "lucky putty" by George Morefield of United Technologies in a March 9, 1984 memo to NASA—was sold by Fuller-O'Brien to Morton-Thiokol.

However, the formula was sold "without warranties," Kelly of Fuller-O'Brien said. Any other supplier who tried to make the material according to the formula could be potentially much more liable to legal action since it lacked the working expertise of Fuller-O'Brien.

The Fuller-O'Brien technology was a time-proven product of trial-and-error. Kelly described how it was formulated: "It was a technology lying there on the workbench, if you like, as are so many others, and someone said, 'Can it do this job?,' and they tried it, and it worked, and then it worked again and so it came into being."

But technical formulas, like the recipes of great chefs, aren't easily replicable. Morton-Thiokol couldn't find another supplier. The main problem was "finding a company that is going to continue making the putty," said Tom Russell, a Morton-Thiokol spokesman.

When that proved to be impossible, Morton-Thiokol began to use a substitute purchased from Randolph Products of Carlstadt, New Jersey. That putty—which also contains asbestos—was based on a "government-supplied formula" for use in jet airplane engines, according to Wendell Randolph, a company official. Jet engines, of course, are subject to much gentler temperature extremes than space rockets. But a "government supplied formula" would be unlikely to provoke lawsuits because it—and its manufacturer—would be protected by the sovereign immunity principle that the U.S. Navy has invoked in denying any legal responsibility for workers exposed to asbestos while building ships during World War II.

It would appear obvious that a chemical analysis of the differences between the two putties could be the key to understanding the exact technical reason why the Challenger accident occurred. "With so much evidence incriminating this putty," Sen. Steve Symms (R - Idaho) wrote James Webb, NASA administrator on June 25, 1986, "the need for chemical analysis should have been immediately obvious."

Symms noted that a February, 1984 NASA memo had warned: "Specifically, concern is raised about the type 2 Randolph zinc chromate (ZCT) putty's sensitivity to humidity and temperature." It replaced the Fullen-O'Brien product referred to by NASA and Morton Thiokol officials as 'lucky putty.'" The senator, a member of the Senate Committee on Environment and Public Works, observed, "Only one-eighth of the [Presidential] commission report is devoted to the immediate cause of the Challenger accident, the failure of the pressure seal in the aft-field joint of the right solid-rocket motor." He went on to observe the commission report had focused on "institutional problems" rather than bearing down on the "numerous unheeded warnings" about the putty's ability to protect the O-rings against superheated exhaust gases.

Symms recognized the Presidential commission was "primarily composed of persons with management and engineering backgrounds. That a chemical analysis of the putty was not a an object of the commission's scrutiny is understandable," he wrote, "but now is the time to address these unanswered questions. . . .

"I encourage you to discover what, if anything, is the functional and chemical difference between the earlier and more recent putties," Symms concluded. In a return letter to Symms, dated August 20, 1986, John F. Murphy, assistant NASA administrator, effectively conceded—in typical bureaucratic language—that the putty had, in fact, been responsible for the crash:

"At the time Fuller O'Brien discontinued manufacturing its vacuum putty product line, which had been qualified on the space shuttle SRM, MTI [Morton-Thiokol, Inc.] initiated the search for a replacement putty. Putty from Randolph Products was chosen based on its prior use in other solid rocket motors and results of bench testing performed at MTI. The bench tests included chemical analyses and physical properties determinations such as thermal conductivity. The Randolph putty had some chemical differences, but it was comparable in most physical aspects, including its performance in the full scale static test motors. Environmental effects on the Randolph putty were more pronounced as evidenced by the data from tests performed to understand the effects of temperature and humidity. Controls were implemented to ship and store the putty in sealed containers and in temperature controlled cold boxes. Out-of-storage time prior to installation on an SRM was also limited."

The letter went on to exculpate all the involved parties: "After all the above laboratory and static firing tests, specification changes were made by MTI and MSFC (NASA's Marshall Space Flight Center) regarding the handling and use of the Randolph putty. It was felt that a satisfactory SRM field joint sealing system had been achieved."

Whatever was felt—rather than known or proven—the bottom line is that the putty was, in fact, the culprit: "As a result of the Challenger accident investigation and STM redesign activities and other technical assessments," Murphy conceded to Symms, "an approach that is generally accepted by all technical managers at NASA and MTI is that any proposed joint redesign for the SRM shall not use putty in the field or nozzle/case joints." The chemical analysis between the two types of putty Symms asked for has not been performed, or, at least, the results have not been made public. But the functional differences between the two are apparent.

In August of 1985, six months before the Challenger explosion, a Titan missile with the same O-ring design using the Randolph putty blew up under similar circumstances. In April of 1986, three months after the Challenger accident, a second Titan exploded, again under similar circumstances.

"Two failures in a row are pretty bad," said Jeffrey Richelson, a military reconnaissance expert at American University in Washington, D.C. "We had 50 successful launches before this." Such overwhelming circumstantial evidence is difficult to ignore. And the refusal to use putty that had worked successfully in 50 launches can only be the bureaucratic equivalent of chutzpah, the gall with which a killer of both parents pleads for the mercy of the court on the grounds he is an orphan.

The most immediate practical effect has been that both the American civilian and military space programs have been effectively, if not deliberately, sabotaged.

A game of Russian roulette, in Feynman's phrase, had been played out—and lost. For it isn't enough to address the technical or, for that matter, the managerial or even the scientific conditions that made the accident not only possible, but inevitable.

The political issues must be addressed.

In technical terms, what had started as erosion of the O-rings allowed a "burnthrough" of hot gasses to the engine tanks caused by the chemical composition of the putty itself or a failure to work in combination with the other elements in the seals. In liftoff, avenues were created for the escape of extremely hot gases that bore down heavily on the O-rings. A backup secondary ring apparently prevented ignition on the earlier launches.

But the real failure occurred because people were not thinking straight. They were not questioning their assumptions, calculating the odds, recognizing that political considerations cannot always be supreme—you can't have it all, success and perfect safety.

Time—and the odds—were running out in the Challenger program. The crash was, in fact, an accident waiting to happen. No one was more aware of that than Roger Boisjoly, a engineer with Morton-Thiokol who first noticed trouble with charred O-rings after a launch in 50-degree weather in January of 1985. By July 1, a full six months before the crash, Boisjoly, after running a series of simulated tests, was concerned enough to write a letter to R.K. Lud, vice president for engineering for Morton-Thiokol. In it, he warned: "This letter is written to ensure that management is fully aware of the seriousness of the current O-ring problem from an engineering standpoint. The mistaken position on the joint problem was to fly without fear of failure, and to run a series of design evaluations which would ultimately lead to a solution or, at least, a significant reduction of the erosion problem.

". . .This position is now drastically changed as a result of the SRM nozzle joint erosion which eroded a secondary O-ring with the primary O-ring never sealing. If the same scenario should occur in a field joint [one actually used in flight], and it could, then it's jump ball as to the success or failure of the joint. . . . The results would be a catastrophe of the highest order—loss of human life."

Boisjoly urged that an informal team which had been investigating the problem "officially be given the responsibility and the authority to execute the work that needs to be done on a non-interference basis [fulltime assignment until completed]. It is my honest and very real fear," he concluded, "that if we do not take immediate action. . .then we stand in jeopardy of losing a flight along with all the launch pad facilities.

Boisjoly's recommendation was accepted and a formal team assembled by August 31. But neither he nor the team received the cooperation needed to do the job properly. On October 4, 1985, he wrote a memo to his superior complaining of a "business-as-usual-attitude" on the part of what were supposed to be supporting organizations.

"[E]ven NASA perceives that the team is being blocked in its tasks," he wrote. "The upper management apparently feels the SRM program is ours, and the customer be dammed."

But if Boisjoly had a hard time with his own management, he experienced even more misery at the hands of NASA officials who had apparently adopted a "routine" policy permitting flights to continue even though the problem was only being perfunctorily addressed and far from being resolved.

The solid fuel rocket division headed by Lawrence Mulloy had taken note of the problem in a monthly report dated July 2, which stated: "The first 12 flights had four occurrences of primary O-ring erosion." A handwritten note on the report noted "better putty" over a recommendation to "apply more concentrated/accelerated activity in this area."

But fears that a "quick fix" approach to the problem would be adopted had already been raised in a memo from Russ Bardos, a NASA official, to David Winterhalter, NASA's chief of propulsion in a July 22 memo. And, in fact, the Presidential commission learned only days before its last session that six waivers were "routinely" signed permitting flights after July of 1985. Mulloy, who signed the waivers characterized them as nothing more than an acknowledgement that a problem existed and needed to be fixed.

But the waiver process specifically required that before each of the six launches after July, 1985, engineers had to positively justify continuation of flights despite the waivers. But the problem was just not serious enough to suspend flights, according to Mulloy. "A lot of people concurred in the decision to keep flying." No one at NASA, Morton-Thiokol or on the commission thought to ask why the putty was changed in the first place. Boisjoly and his team members also had to contend with a problem in logic all too familiar with anyone who has tried to reason with regulatory agencies and the environmental lobby about the relative risks associated with exposure to a carcinogen–proving a negative.

As a classic case in point, Boisjoly at a meeting with NASA officials the night before the Challenger launch found himself in a Catch-22 position:

"It was up to us to prove beyond a shadow of a doubt that it was not safe to launch. This is the total reverse to what the position usually is in a pre-flight conversation. We were put into a position to prove we could not launch rather than being put into a position to prove we have enough data to launch."

Boisjoly was questioned about the pre-launch meeting by Feynman: "I take it you were trying to find proof the seal would fail," Feynman said. "And, of course, you didn't and had you proved (theoretically) that it would fail, you would have found yourself incorrect because five didn't fail."

"That is correct," Boisjoly said. "I was very concerned that timing would place us in another regime; and that was the whole basis of my fighting that night."

Obviously such a logical process–proving a negative–is "flawed," as the Presidential commission observed. Normally, as Feynman pointed out, in pre-launch flight reviews officials "should agonize whether they can go," if the seals had eroded on the previous flight or there was any other indicator they were pushing their luck. But, with each successful flight NASA and Morton-Thiokol officials got cocky–wearing "management" rather than "engineering" hats– and lowered their standards "because they got away with it," Feynman wrote. That is the fundamental flaw in the thought processes behind the regulation of carcinogens: a basic failure to assess relative risks. The assumption is if you can find a carcinogen, ban it, without any thought of what the consequences might be. Don't even try to make any determination whether adequate substitutes are available, and whether they, too, might be carcinogenic–or, as in the case of the

Challenger, much less effective in preventing accidents.

But the brute fact is we live in a world in which there are few easy alternatives. There's as much scientific wisdom as common sense in the old saying, "Better the devil you know than the devil you don't know."

"All the substitutes for asbestos are, in fact carcinogenic," according to Dr. Malcolm Ross of the U.S. Geological Survey. "Ceramic fibers, rock wool, fiberglass, all have been found carcinogenic in laboratory animals."

Boisjoly couldn't prove the shuttle would blow up. The scientists can't prove that given some unique set of circumstances a single asbestos fiber can kill. They can prove in every case of chrysotile-related disease exposure had been massive, and in 80 out of 81 cases occurred in cigarette smokers. They can't point to a single case associated with non-occupational exposure. There might be a case out there, but none has been—or can be, as a practical matter—found. After all, dragons could exist—just try to find one.

All that Boisjoly could prove, on the basis of mounting and incontrovertible evidence, was that the odds were mounting too high to risk another launch. His was the impeccable logic of mathematics and science. It was not a willingness to gamble with life and death that has characterized so many regulatory bans of highly useful and valuable substances, such as asbestos, on the assumption that something will show up to replace it.

That is Russian roulette, and it could be played on millions of cars that, up until now, have had assured braking power using asbestos formulas developed in much the same manner as the Fuller-O'Brien putty. And Ross' fears that the substitutes for asbestos may be even more dangerous is almost universally shared in the scientific community.

"Ceramic fiber and aramid fiber," Dr. J.M.G. Davis, head of the pathology department at the Institute of Occupational Medicine in Edinburgh, has written, "appear to be extremely durable. . .much more durable than chrysotile" [white asbestos, the most widely used type]." But he then warns: "It is likely that they will only be safer than chrysotile in industrial use if they can be handled with the production of far fewer respirable fibers than chrysotile."

If there is a social lesson to be learned from the Challenger crash that transcends the narrow question of who did what when, it is perhaps best expressed in the words of Dr. Robert Murray, president of the International Occupational Medicine Association, commenting on EPA's proposed asbestos ban:

"The lesson is we ban the uses that present the greatest hazards, but retain those where the hazard can be controlled. Rather than 'unreasonable risk,' I believe the EPA proposal to be based on 'unreasonable fear' supported by dubious evidence and special pleading such as always to appeal to the natural fears of people to be rid of what they conceive to be an insidious and sinister threat to their lives.

"Paranoia in an individual is a serious form of psychosis. As with other psychoses, it has a social force as well, a sort of collective paranoia. Here the false premise is fanned by public opinion until it becomes an obsession. This was true in ancient times with witchcraft. It is even more dangerous today when false prophets have the opportunity of disseminating their opinions more widely through television and newspapers and playing on the credulity of people.

"In the emotional hysteria that results, panic actions are taken which are regretted in the light of hindsight, just as the good burghers of Salem in remorse tried to compensate the victims of precipitate action."

The United States government has acknowledged its guilt for the victims who died because of a lack of asbestos by paying the families of the Challenger astronauts substantial settlements. Ironically, it has never accepted responsibility for the much greater number of lives lost to excessive asbestos exposure in World War II.

The Challenger crash wasn't an appointment in Samarra, but an accident waiting to happen because too many people failed to understand that, taken far enough, concern over safety and health can kill.

Unfortunately, EPA has not taken the lesson of the Challenger into account in announcing its total ban on asbestos in the summer of 1989, a ban the agency itself admits would save, at best, seven to ten lives every year for the next 15 years.

The lesson of the Challenger as well as Murray's advice went unheeded at EPA: "I suggest EPA pause and consider the situation soberly before embarking on such an unjustified course of action."

But NASA did pay attention.

An independent three-member redesign panel appointed by the National Research Council at the recommendation of the Rogers Commission replaced the zinc chromate seal with two asbestos-silicate seals.

The three members of the panel were Jack L. Blumenthal, chief engineer of engineering operations of TRW Space & Technology; Robert C. Anderson, who worked on the Minuteman missile and Apollo program for TRW; and H. Guyford Stever. Stever, the chairman of the panel, was on the faculty of Massachusetts Institute of Technology (MIT) for 20 years, where he was dean of mechanical engineering. He also served as president of Carnegie-Mellon University, director of the National Science Foundation, and science advisor to President Gerald Ford.

The panel confirmed, according to a co-authored article in *CHEMTECH*, November, 1989, that the "slow dynamic response (resiliency). . .of the O-ring seals at low temperature" was responsible for the crash.

The original Fuller-O'Brien putty was "so plastic I could stick my fingers in it," Malcolm Ross of the U.S. Geological Survey has said. "By contrast, the zinc chromate was stiff even at high temperatures."

No future members of space missions should be killed because of a lack of asbestos in O-ring seals.

But that's no guarantee that other tragedies may not yet be in the making.

The king may not do any wrong, but don't tell that to his victims, for whom the Challenger disaster was a dismal debacle in the cause of a sordid racket.

As a business advisor, an accountant should think like an entrepreneur. So why did Stuart Seltzer, a 24-year-old auditor for Arthur Anderson & Co lose his job for doing that? Seltzer and two friends raised enough money to shuttle 450 pounds of the Berlin Wall to New York, where they were cut into chunks. . .for sale at $5, $10 and $20, making an initial $20,000. The Arthur Anderson personnel director objected, saying "the rock might contain asbestos, which could harm people. . .and people who bought them would sue Arthur Anderson as the 'deep pocket.'" Seltzer was given a choice between his "enterprise" and employment at Anderson. He decided to quit. . . .
"I like to take chances," he said.
—Wall Street Journal, Dec. 19, 1989

6 Miracle Mineral or Fatal Fiber? ■■■■

Asbestos was the first mineral whose uses were entirely peaceful. Unlike the instruments for war and hunting produced by the Bronze, Iron, and Copper Ages which preceded its introduction into the life of early societies, asbestos was always used to protect and enhance human life. That was enough, in itself, to warrant the label attached to it for hundreds of years—the "miracle mineral."

It was originally called a "miracle mineral" because its fibers, when pulverized out of rock, twisted into a wick and doused in oil, would burn without being destroyed. The fifth century B.C. Athenian sculptor Calimachus used asbestos in the wick of a gold lamp held by a gold statute of the goddess Athena. The Emperor Charlemagne (768-814 B.C.) would amaze and amuse his guests by having asbestos tablecloths placed over fires to burn away food crumbs and grease and returned to the table clean and crisp—if a trifle hot—for another course.

The fireproofing and insulating qualities of asbestos have been long known. Finnish peasants, as far back as 2500 B.C., used asbestos fire-proofed cookware and also stuffed it between chinks in log cabins for insulation. Bushmanoid pottery, dating back to the stone age in the southern Sudan of Africa, contains asbestos.

Plutarch (46-120 A.D.) wrote of rocks yielding "soft, petrous filaments" which are woven into "towels, nets and women's hair coverings which cannot be burned by fire; but, if any become soiled by use, their owners throw them into a blazing fire and take them out bright and clean."

Asbestos, indeed, until the 1960s was considered a great life saver rather than life taker, a source of protection rather than danger.

Asbestos was used in body armor in the Middle Ages, and in clothes for firemen in early 19th century Italy. A mill was set up near Rome by Pope Pius IX in 1830 to produce paper that would protect Papal Bulls and other documents from fire.

The art of weaving asbestos was kept secret for many centuries, but, as first publicly revealed, in the late 17th century made possible by first boiling the pulverized mineral in strong oil. The principal reason asbestos did not come into much wider use well until the 19th century seems to be the fact that until the Industrial Revolution, and particularly the introduction of the steam engine, there was no great need for insulation against high temperatures. Further, the types of asbestos found in Europe, other than the Ural Mountains, were not well adapted for weaving and incorporation into other materials as varied as clothing and cement.

There are two basic varieties of asbestos, serpentine or chrysotile and the amphiboles. Serpentine is a rock composed of up to five percent chrysotile or "white" asbestos fibers, which are considerably thinner than the amphibole varieties, principally amosite or "brown" and crocidolite or "blue" fibers.

The principal deposits of chrysotile are in the Soviet Union and North America, primarily Canada. The discovery of chrysotile deposits in the Urals under Peter the Great around 1720 made possible an industry in textiles, socks, gloves and handbags that flourished for several decades. The first mention of North American chrysotile, according to Dr. Irving J. Selikoff and Douglas H.K. Lee in *Asbestos and Disease*, was in Benjamin Franklin's *Autobiography*. Shortly after his arrival in London in 1725, Franklin wrote to Dr.

Hans Sloane that he had brought with him "from the northern parts of America...a purse made of the stone asbestos." Franklin later sold it to Sir Hans, who bequeathed his books and curiosities to the Crown as the nucleus of what was to become the British Museum.

Asbestos was still very much a curiosity rather than a commercial product, despite the efforts of a number of Italian manufacturers to make asbestos paper and cloth in the early 19th century. But following the display of a specimen of Canadian asbestos at a London exhibition in 1862, the first company to exploit the mass use of asbestos, the Patent Asbestos Manufacturing Company, was formed in Glasgow, Scotland.

The primary source of asbestos for the company and others to follow was chrysotile from Canada. The first deposit of chrysotile was noted in the Thetford Hills of Quebec in 1847 by Sir William Logan, the initial director of the Geological Survey of Canada. By 1876, an entrepreneur, Andrew Johnson who had noticed the "wooly rock" as a boy, had opened Johnson's Company in Thetford, the pioneer producer of Canadian asbestos. First year production was only 50 tons, but four years later it was six times greater, 300 tons. By the turn of the century, production was 21,408 tons.

With the advent of the arms race of the early 20th century, and the building of more and more massive "Dreadnought" cruisers and destroyers by Germany, England, Japan and the United States, the demand for the mineral multiplied. Asbestos had first been used commercially in Corsica to make stoves because of its insulating qualities. It didn't take naval engineers long to realize those qualities were even more essential in the huge coal-burning ovens producing the steam to drive the "Dreadnoughts"—and save the lives of the men on board if they were hit by enemy fire.

Production quintupled between 1900 and 1911 to 102,224 tons. By the end of the World War I hostilities, production had reached 179,891 tons, dipped sharply to 87,475 tons in 1921 and then climbed, along with the prosperity of the period, to a high of 306,055 tons in 1929. Another decline, reflecting economic conditions, occurred during the '30s, until war again drove production levels from 345,472 tons in 1940 to 558,181 in 1946.

Another set of production and consumption figures were, however, to prove more significant in the long run, the amount of crocidolite or "blue" asbestos primarily, but also amosite or "brown" asbestos imported into this country during the war years. Crocidolite was first discovered in South Africa as early as 1815, but didn't come into general production until the 1920s and '30s.

"Although South Africa remains the major producer of croci-dolite, the actual tonnage is small," Selikoff wrote in *Asbestos and Disease.* "Its production of amosite is somewhat less."

But that was just enough to make the deadly difference.

Dr. Malcolm Ross and other scientists have established that deadly difference in recent years. Their evidence has superseded earlier attempts by Selikoff—eagerly, if not fanatically seconded by the environmental movement and EPA—to indict all forms of asbestos, chrysotile, crocidolite, and amosite alike, of having equally harmful effects. Ross and others have also demonstrated cigarette smoking is massively implicated in asbestos-related disease.

Coincidentally, commercial mining of crocidolite and amosite began in the same period in which the cigarette rolling machine was chartered. "They have walked arm-in-arm through history since then," said Dr. Robert N. Sawyer, a leading asbestos control authority.

Questions about the health effects of asbestos—and ciga-rettes—began to emerge in the 1920s. In 1928, production of crocidolite had only reached 7,000 tons. However, between 1940 and 1976, 367,547 tons were imported into the United States. That is only a fraction of the chrysotile total, but a fatal one, as Ross pointed out. He cited two World War II studies as proof.

In the first, 176 men and women had manufactured or handled crocidolite. Fifty six had died by 1983, eight of lung cancer and nine of mesothelioma (a rare disease of the pleural cavity of the lungs almost invariably associated with asbestos).

In the second, two groups of women were studied, 727 who had only been exposed to crocidolite, and 102 who had worked only with chrysotile. None of those exposed to chrysotile had died by the time the study was concluded, but there were 11 lung cancer deaths and 16 mesothelioma deaths among those exposed only to crocidolite.

Distinctions have to be drawn; differences taken into account. The studies cited, Ross himself agrees, cannot be claimed to be definitive. The cohorts studied were exposed over widely differing periods of time, from as little to five months to as many as four and a half years.

But scientific truth doesn't emerge in blinding revelations. Ross' rebuttal to Selikoff's original studies was no more to be automatically accepted than Selikoff's original findings.

Both men have been awarded recognition by their peers. Selikoff was awarded the Albert Lasker Award in 1955, the most distinguished awarded for medical research into cancer. Ross was the first recipient of the public service award of the Mineralogical Society of America in 1990.

There is a greater burden of evidence to be met than the citation of awards, no matter how distinguished. Even the great discoveries of Newton, Einstein, and Galileo are the product of years of study followed by even more years of review, analysis—and cross-examination—by respected scientific peers. Charles Darwin deferred publication of *The Origin of the Species* for almost 20 years in order to make sure all his critics were satisfied about his methodology, if not his conclusions.

The demands of both the environmental movement and the news media's insistence for simple answers has eroded, if not destroyed, that higher command of science. But the history of research into asbestos has provided a classic example of the dangers, socially and economically, of ignoring that higher command—even one that Selikoff himself now respects.

For Ross's research has been vindicated and Selikoff's warnings found to be comparable to Chicken Little's warnings about the falling of the sky.

While some concern about the health effects of asbestos was

noted as early as the first century A.D. by Pliny the Elder, it wasn't until 1906 that asbestos was linked with pulmonary disease. H. Montagu-Murray reported to the British Departmental Committee on Industrial Diseases a case of fibrosis—scarring—of the lungs of an asbestos worker. No account of the case was apparently published in the scientific literature although Lady Inspector of Factories Adelaide Anderson had tried to include asbestos in a list of dangerous dusts in 1902.

Over the next 20 years, researchers in France, Italy and Canada reported similar cases, but they were associated with tuberculosis. Asbestos, like silica, was assumed to act as an aggravating response to the bacillus which caused tuberculosis.

The turning point in recognition of asbestos as a specific disease agent came with the publication of a 1927 paper, "Pulmonary Asbestosis," in the *British Journal of Medicine* by W.E. Cooke. Earlier Cooke studies had reached the same conclusion. But, in coining a name for the lung scarring linked directly to the materials, "asbestosis," recognition of what was perceived to be a distinctly new disease—associated with workplace exposure—was achieved. Similar word coinages, black lung associated with coal dust and byonossis attributed to cotton dust, were also to become important symbols in the struggle to improve working conditions in mines and factories between 1930 and 1980.

A new scientific and medical specialty was also emerging at the time, occupational medicine. Its practitioners soon were accumulating more evidence that the dust in asbestos mines and mills—often thick enough to distort vision—should be controlled. And by the 1930s, England and Germany had pioneered industrial regulation of asbestos.

Perception of the health problem posed by asbestos was to take another 30 years in the United States. Selikoff, the American scientist who was to produce the studies that led to EPA's successive measures to control and ultimately ban the mineral, was still in medical school in Glasgow. The context of protest against social injustice in which many occupational health studies and programs were—and are—carried out is underlined by the fact that Selikoff had to seek his medical degree overseas because a then-prevailing ethnic quota system prevented him from enrolling in medical schools in the United States before World War II.

But, in the intervening years, now that his seminal studies on asbestos have been questioned, so, too, have his medical credentials.

Selikoff has claimed at various times in his career, in such biographical reference works as *Who's Who* and *American Men and Women of Science*, to have earned his medical degree in both Great Britain and Australia, half-way across a war-torn world, in the same year, 1941.

The two medical faculties cited by Selikoff, the University of Melbourne and the Royal Colleges of Scotland, deny issuing such a degree in 1941.

My correspondence from the two faculties reads as follows:

University of Melbourne:

"Our records show it is correct that Dr. Selikoff was enrolled in this University in 1940 but did not continue his studies. We have no record of any degree(s) or other award conferred on him by this university."

The letter, dated Feb. 20, 1989, was signed by F. Baines for J.B. Potter, Registrar.

Royal College of Surgeons of England:

"I am enclosing an insert from the U.K. Directory and the edited entry shows that Dr. Selikoff was granted the basic medical qualification LRCP/LRCS in 1945 (a co-joint qualification from the Royal College of Physicians of Edinburgh and the Royal College of Surgeons of Edinburgh). He was also granted a similar qualification LRFPS Glasg., by the Royal College of Physicians and Surgeons, and these diplomas would be the equivalent of the M.D. awarded in the U.S.A. I would hardly think that Dr. Selikoff would have been awarded a doctorate in medicine in 1941, four years before he was awarded the basic medical qualifications in 1945."

A handwritten note was on the back of the Royal College letter, dated Jan. 16, 1989, and signed by E.G. Bruce, record officer:

"Please note the doctorate of medicine in the British Isles is a much higher degree. A medical practitioner cannot apply for this qualification at least 5 years since gaining basic medical qualifications and then has to be approved by an examining board."

But, apparently without the basic British medical qualification, never mind the advanced degree, Selikoff reported himself as serving during 1941 and 1945 as a fellow in pathology at Mt. Sinai Hospital in New York, as an intern at Newark, N.J. Hospital and as a resident at Sea View Hospital in New York.

"If the British records are correct," *The Washington Times* reported on April 21, 1989, "he participated in these programs without the required doctorate"—British or American.

Selikoff's own explanation, when contacted, was only, "It was the war."

The Washington Times article concluded:

"In short, if the epidemiological evidence against Dr. Selikoff's work isn't enough to persuade bureaucrats to rethink our asbestos policy, perhaps the dispute about his credentials is."

The dispute hasn't done so any more than previous controversies over Selikoff.

He was described, for example, years earlier in the *Boston Globe Magazine* of January 16, 1983 as "not only a gadfly but a zealot. . . . His work is said to be invaluable to present knowledge of asbestos hazards, but he is regarded by some experts in asbestos disease as a zealot whose careless public statements about the dangers of asbestos have contributed to public paranoia. Selikoff, for his part, has implied that several of the authorities mentioned in this article are running dogs of industry."

And, indeed, asbestos mining in Canada, in the past, was also associated with political and social injustice. The mines were owned primarily by American and British companies and seen as a symbol of economic colonialism. A general strike against the mines in the late '40s led to the creation of the Parti Quebecquois, the separatist movement in the province, and the eventual election of Pierre Trudeau, one of the leaders of the strike, as prime minister of Canada itself.

Ironically, 30 years later, with the ownership of the mines vested largely in private and public Canadian hands, the Confederation of Canadian Labor, along with the Quebec and national government, is vehemently protesting EPA's proposed ban on asbestos. Environmental imperialism has apparently replaced economic colonialism, and the old question of balancing jobs and worker safety is taking a new form. Cost/benefit analysis takes many forms, most guided by the realities of the moment.

On the other hand, there is little debate that the questions raised in the United States—at least until the '70s—about asbestos had been decided with little consideration for worker health.

The U.S. Department of Labor, acting under the Walsh-Healey Act of 1938, did not get around to setting a permissible exposure limit (PEL) to asbestos in the workplace until 1969. Then, it set the limit at 12 fibers per cubic centimeter (f/cc), although the U.S. Public Health Service recommended 5 f/cc in 1938. The Occupational Safety and Health Administration (OSHA), shortly after it came into existence in 1971 cut the PEL to 5 f/cc, and later reduced it to 2 f/cc in 1976. A further PEL exposure limit 2 f/cc to 0.2 fibers—about the least amount detectable with advanced electron microscopes—was imposed by OSHA in 1986.

There can be no dispute that the U.S. Government, as well as the asbestos industry, was negligent in failing to regulate exposure to asbestos much earlier. The negligence, in fact, may even have been a matter of public policy dictated directly from the White House. A letter dated March 11, 1941, from Cmdr. C. S. Stephenson USN, officer-in-charge of preventive medicine for the Navy concedes serious health problems existed with asbestos: "Asbestosis: We are having a considerable amount of work done with asbestos and from my observations we are not protecting the men as we should. This is a matter of official report from several of our navy yards."

Nevertheless, Stephenson, in the letter retrieved from the National Archives, justified a policy of refusing to allow inspectors from the Labor Department and the U. S. Public Health Service to inspect working conditions in yards directly or indirectly run by the Navy. Two reaons were given: 1. The Navy already "had medical officers in the yards," and 2. "President Roosevelt thought this might not be the best policy, due to the fact they [the Labor Department and Public Health inspectors] might cause disturbance in the labor element."

Negligence, however, is consistent with an absolutist pattern in American political life. Problems are ignored until action is obviously needed, but then extreme—indeed excessive—action is taken. Until the advent of the environmental movement that tendency, dubbed "the paranoid tradition in American politics," by historian Richard Hofstader, had found its greatest expression in Prohibition.

In assessing public policy on asbestos, it should be noted that the reaction has been similarly excessive and out of proportion to the problem. For example, a successful ban on liquor could save at least 100,000 lives a year. A similar ban on asbestos could save, at best, according to EPA statistics, a maximum of 126 lives a year. And an asbestos ban could potentially cause thousands of deaths from ineffective braking systems and fires in buildings.

In assessing the dangers associated with asbestos, it should be recognized that asbestosis—by far the most common disease associated with asbestos—although it was identified as a specific disease as early as 1920, is very difficult to diagnose unless there is a significant reduction in lung function. If any obstruction of the lungs is present, emphysema, for example, the cause can almost always be attributed to cigarette smoking.

Asbestosis is not a tumor or a viral or a bacteria-caused ailment. It is a reaction of lung tissue to the presence of mineral fibers, and can only be diagnosed with 100 percent certainty by biopsies. Chest X-rays, however, have been the means usually used to diagnose asbestosis, particularly by Selikoff and his associates.

X-rays may show white lines indicating abnormalities, but the lines usually look the same as those found in a variety of other diseases. X-ray interpretation in cases of slight or early asbestosis has been called "the most subjective in medicine" by Dr. Raymond Murphy, an occupational lung specialist at Boston's Faulkner Hospital.

Dr. Barry Levine, a lung specialist at Harvard Medical School, agrees: "It's a rare phenomenon to find true asbestotic disease. . . . A person with obstructive lung disease (a cigarette smoker's disease such as emphysema) can be read as having an X-ray comparable with asbestosis."

Asbestosis is also associated with high levels of exposure in the past. "I have not seen a new case of asbestosis in two or three years," said Dr. Edward A. Gaensler, who heads a national asbestos research project funded by the National Institutes of Health. "Ninety percent of the cases referred to this laboratory do not have asbestosis. Asbestosis is a disappearing disease."

Diagnosis becomes even more complicated when asbestos is implicated in cases of lung cancer and mesothelioma, a rare disease of the lining of the lung. It wasn't until 1960 that a paper published by J.C. Wagner of the Pneumicosis Institute in Cardiff, Wales demonstrated a connection between asbestos and mesothelioma. However, mesothelioma is now believed to have a "strong genetic component," according to Edith Efron, author of *The Apocalyptics: Cancer and the Big Lie.*

But despite the scientific doubts and uncertainties, the single fiber theory first officially declared by Russell Peterson, chairman of the Council on Environmental Quality, in 1976 Congressional hearings on the then-proposed Toxic Substances Act, had become political dogma: "No level of asbestos fibers in tissues can be regarded as safe—a single fiber may initiate a response."

The discovery of nature, of the ways of planets, and plants and animals, required first the conquest of common sense. Science would advance not by authenticating everyday experience but by grasping paradox, adventuring into the unknown. . .seek[ing] clues to the mystery of an ever-changing nature.
—Daniel J. Boorstin

7 Studying Asbestos

Any appreciation of value of asbestos studies must begin with recognition of the fact that the first major study was conducted by Sir Richard Doll of Oxford University in the 1950s. Doll studied asbestos workers with at least 20 years of exposure to the mineral in textile plants, and found they were experiencing 10 times the lung cancer deaths of non-asbestos workers. But he also found cigarette smoking multiplied the risk many times over, and concentrated then on now internationally honored epidemiological studies conclusively proving cigarette smoking to be the primary cause of lung cancer.

Dr. Irving Selikoff took up studies of asbestos as a prime cause of disease after Doll had shifted his focus to cigarette smoking. That was to make Selikoff a hero of the media, labor unions and the environmental movement—but highly controversial in the scientific community. In the '60s, Selikoff tracked down many of the workers who had been carried on the rolls of an asbestos workers union, and found by 1973 the group had experienced a death rate 50% greater than the average white male.

In retrospect, those studies were seriously flawed, even primitive, by current scientific standards. To begin with, the comparison to the average white male provided a totally inadequate "control"—a group sufficiently dissimilar in socio-cultural characteristics so that any sharp difference in health could not be related to a single factor, in this case asbestos, under study. Even in the '40s and '50s, white collar workers were much less likely to be smokers than blue collar

101

workers, and, of course, were not subject to occupational exposure from heavy concentrations of dusts.

Moreover, the general health of the average white male was much more likely to be better, because of access to better nutrition and medical attention. That was particularly true during the Depression years. Further, almost all World War II workers had either been too old to be drafted, or even more significant, unable to pass the physical for admission to the armed forces.

Another difficulty with the Selikoff studies, little appreciated at the time, but seen now as very disturbing, was the failure to assemble complete cohorts. Those are groups with all the workers located in a particular workplace, and tracked for 20 years or more. Almost one-third of the Selikoff cohorts were unaccounted for 20 years later. Moreover, the Selikoff studies assumed the three different types of asbestos, chrysotile, crocidolite and amosite, although different in their mineral characteristics as well as the length and diameter of the individual fibers, would have the same biological effects in humans.

That was one of two great weaknesses in the studies, first pointed out by Ross, and subsequently confirmed by other scientists.

The second was one Selikoff conceded from the outset—but which was virtually ignored in the subsequent debates about the dangers posed by asbestos. The Selikoff studies, again from the outset, made the central point that asbestos while dangerous, was infinitely more so in combination with cigarette smoking, Selikoff published studies in the '60s which estimated smoking increased risks as 80 to 90 times. "[H]ad it not been for cigarette smoking, many of these risks would have been avoided," Selikoff wrote in one paper.

However, Selikoff's work appeared at a time when the New Left and the Old Left had joined, followed by the media, against a society which could send young men to die in a useless war, impose quotas on some ethnic groups and deny affirmative action to others, even poison the places here we work and play—our environment—with carcinogens. It was a world view, a zeitgeist that was impatient with "ifs" and "buts," that disdained caveats, such as the influence of smoking on asbestos-related diseases, as "sell-outs" to capitalist interests when "everyone knew" cancer was a by-product of industrial civilization.

And Selikoff was about to become a media star, first in a lengthy article in *The New Yorker* by Paul Brodeur, then in frequent appearances on national TV programs, always in a crisp, white laboratory coat. Selikoff gave every indication of loving the role. An article in *The Boston Globe* of Jan. 16, 1983 described him this way:

"Selikoff is not only a scholar but a gadfly. He is aggressive, accessible to the press, always good for a snappy quote, and he offers higher estimates of risk for most asbestos-related disease than any other investigators. His work is said to be invaluable to present knowledge of asbestos hazards, but he is regarded by some experts as a zealot whose careless public statements about the dangers of asbestos have contributed to public paranoia. Selikoff, for his part, has implied several of the authorities mentioned in this *Globe* article are running dogs of the asbestos industry."

Dr. J.C. McDonald of McGill University and Dr. Hans Weill of Tulane University, were among Selikoff's critics then and now. Weill, for example, is quoted in the *Globe* article as saying: "There is general agreement except for the New York group [Selikoff and his associates] there is no such thing as pleural asbestosis."

But Brodeur, in two more sets of articles in *The New Yorker* and three books, cavalierly dismissed the findings of any scientists who disagreed with Selikoff. He implied their studies were not sufficiently balanced because the research was funded indirectly by industry as well as government institutions and some unions. Until the *Globe* article appeared, no such similar criticism was directed at Selikoff.

Yet "Selikoff and his laboratory staff go around the country conducting disease screenings in union halls," according to the *Globe* article, "reporting. . .vague and ominous findings. . .to the workers, the unions and often the press. No medical or occupational records, from either company or union, were used."

But the national news media in Washington and New York does not read even such stalwart establishment papers as the *Globe*. They are more likely to rely on *The New Yorker* to do their original research because the magazine has an apolitical reputation for detailed reporting and enjoys enormous credibility among journalists. But, in reality, *The New Yorker* has been the environmental movement's most fervent—and unquestioning—advocate in the media ever since it published Rachel Carson's *Silent Spring*, the movement's call to arms, in its entirety in 1962, even before it was published in book form.

Reporting that seeks to influence events and shape public policy is often more effective when it omits information rather than when it openly advocates positions. "Never let the facts stand in the way of a good story," is an old adage of newspaper rewrite men. Brodeur, in his articles and books, not only attacked McDonald and Weill, but failed to mention cigarette smoking. Yet even Selikoff pointed out, in a 1969 article in the *Journal of the American Medical Association:*

"Asbestos exposure alone is not the entire explanation. Calculations suggest that asbestos workers who smoke have about 92 times the risk of dying from broncogenic carcinoma [lung cancer] than men who neither work with asbestos nor smoke."

Unfortunately, *The New Yorker's* much vaunted team of "fact checkers," whose work presumably verifies every word that's printed, cannot account for facts that are omitted. And Brodeur omitted the crucial facts about cigarette smoking.

I sent a letter to William Shawn, then editor of *The New Yorker* in February of 1986, complaining:

"There is a consistent pattern here, one that, at best, misdirects the public and, at worst, deceives it. And both should be a matter of concern to anyone concerned about the credibility of the media. That pattern was evident in Brodeur's series on asbestos printed in *The New Yorker* last summer. Absolutely no mention whatsoever was made of the official report to the British Ministry of Health commissioned from Sir Richard Doll. The omission of Sir Richard, not only in that series, but in Brodeur's earlier writings is particularly significant. Sir Richard, after all, is the acknowledged world authority on the relationship between smoking and lung cancer, and his asbestos studies predate and are fully as extensive as those of [Dr. Irving J.] Selikoff, Brodeur's primary scientific source of information.

"Even more extraordinary," my letter continued, "is the fact that Brodeur could write so extensively about asbestos over the years and never mention cigarette smoking. Even Selikoff in his most widely cited study, said: 'had it not been for cigarette smoking, many of the excess deaths would have been avoided.'"

Significant portions of my letter follow:

"Indeed, the recently released Surgeon General's report on cigarette smoking in the workplace makes the following central observation in the preface: 'Smoking and occupational exposures can interact synergistically to create more disease than the sum of the separate exposures. This kind of interaction is exemplified by the relationship between asbestos exposure and smoking. . . . In other words, for those workers who both smoke and are exposed to asbestos, the risk of developing and dying from lung cancer is 5,000 percent greater than the risk for individuals who neither smoke nor are exposed.'

"Obviously, Brodeur cannot be faulted for failing to quote a study [the Surgeon-General's report] which had not been published at the time he wrote his articles. But he can be criticized—and severely—for adopting an openly partisan attitude on questions he, by self-admission, lacks the technical competence to judge; and interjecting himself into public hearings he was ostensibly present to report, not to influence.

"The best example of this partisan attitude is his own account in the Nov. 19, 1973 issue of The New Yorker of a OSHA public hearing. One of the principal witnesses, Dr. J.C. McDonald, an internationally respected epidemiologist, testified there were significant differences between the health effects of different types of asbestos. White asbestos, the kind used in buildings, he said, did not seem to have the same adverse effects as the blue and brown asbestos used by the shipyard workers studied by Selikoff.

"'As a layman,' Brodeur wrote in his account, 'I had little way of judging the scientific validity of Dr. McDonald's work except through the observations of those members of the independent medical community who had communicated their opinions of it to me."

"Independent" is the adjective scientists allied with the environmental movement have chosen to describe themselves. For some reason, journalists who are pains to describe the much more broadly based organization, Common Cause, as "a self-styled citizens lobby," have uncritically accepted that self-serving description of what others have called "apocalyptic" scientists without comment. These so-called "independent" scientists have not hesitated to resort to blatantly political attacks on scientists with the temerity or independence to disagree with them on issues. And the attacks have had very little, if anything, to do with the scientific merits of a case.

This instance was to prove no exception to the rule, with one significant exception: The "independent scientists" got an "independent" reporter, Brodeur, to do their hatchet work for them.

According to Brodeur, "Dr. McDonald prefaced his testimony by noting he was a 'fulltime' employee [as chairman of the department of epidemiology and health] at McGill University, and an independent research worker. McDonald said: 'I do not work with, nor am I associated with any asbestos producer or manufacturer. The research I shall be describing is supported by grants, not to me, but to McGill University, from a number of sources—The Institute of Occupational Safety and Health, the Canadian government, the British Medical Service and the United States Public Health Service. I am not here to support the testimony of the Johns-Manville [now the Manville] Corporation or any other body affected by the proposed regulation.'"

Johns-Manville and other asbestos companies were opposing an OSHA proposal calling for a maximum exposure level of two fibers per cubic centimeter in the workplace as impractical. McDonald, speaking as a scientist, believed in Brodeur's words, "that a reasonable standard for chrysotile [white] mines and mills would be somewhere between five and nine fibers per cubic centimeter.

McDonald was cross-examined by Dr. William Nicholson, a close associate of Selikoff, and Nicholas DeGregorio of the Department of

Labor. Neither were able to shake McDonald on any scientific point. DeGregorio was reduced to protesting that any dispute of Selikoff's findings amounted to scientific lese majeste. In Brodeur's words, De-Gregorio commented "with some asperity he had never heard of the validity of Dr. Selikoff's. . .study of the asbestos insulation workers questioned by any of the leading epidemiologists in the field."

But McDonald was, in fact, a leading epidemiologist and his testimony was now in record—and, for all practical purposes, uncontested.

It was at this point the self-described layman, Brodeur, intervened in the proceedings using guilt-by-association tactics: "[S]ince anyone attending the public hearings had the right to cross-examine witnesses, including members of the press, I decided to ask him some questions," Brodeur wrote.   Members of the media rarely avail themselves of such a right, preferring to reserve their queries for formal press conferences or private interviews, rather than intervening in hearings where public policies are determined.   Brodeur's questions brought out the fact that one of the four specific funding sources for McDonald's work, the Institute of Occupational Safety and Health was an affiliate of the Quebec Mining Association.

"When I took my seat," Brodeur reported, "Dr. McDonald had just indirectly admitted that Johns Manville. . .had helped support his study. . .It seemed unnecessary to point out to the representatives of industry, labor, government, and the independent scientific and media community something that many of them knew—that Johns-Manville is, and has been, for a quarter of a century, the dominant member of the Quebec Mining Association."

But Johns-Manville is not now and never has been dominant in the affairs of the Canadian government, the British Medical Research Council, the U.S. Public Health Service or McGill University. That would probably have come out if Brodeur hadn't sat down, having sown his seed of innuendo. That might well have come out had Brodeur himself been subject to cross examination.

I had prefaced my complaint by noting that, as a journalist, I respect *The New Yorker:* "I feel like a monk accusing the Pope of heresy." I noted that my comments had been published, in different form, in a number of highly reputable mainstream publications, such as *The Detroit News* where my 1985 series on asbestos had won the paper's top prize of the year and been nominated for a Pulitzer. I then, as a reporter, directed some very serious allegations against Shawn as an editor:

"You have allowed, if not encouraged, writers on environmental issues to abandon traditional journalistic standards of fairness, accuracy and thoroughness to turn the most respected reporting medium in the country into a blatantly one-sided propagandist for disputed—indeed, almost entirely discredited—scientific theories. You have either deliberately misinformed your readers, or have been gullible enough to be misinformed by your writers. In either case, your professional competence and that of *The New Yorker* stand in question."

The letter made specific reference to Brodeur's conduct at the OSHA hearing, and asked: "Do you support writers who engage in character assassination in the name of ecology, guilt by association for the sake of the environment, McCarthyism in the cause of Greenpeace?"

There was no answer, not even an acknowledgement. Brodeur agreed to—and then backed out of—a debate with me on National Public Radio. And, of course, there was no debate in *The New Yorker*. *The New Yorker* does not publish editorials—or letters to the editor—it finds them improper, somewhat like questioning the Pope. Rather, *The New Yorker* shapes public opinion by ignoring those facts it finds inconvenient and refusing to print letters from those who point out the omissions. An exceptionally long 16-page "Department of Amplification" response to criticism of a Brodeur article on cancer presumably caused by electric power line fields did appear in *The New Yorker* November 19, 1990. It was written, however, by Brodeur, not his critics. *The New Yorker* in its omniscience, consigns to a black hole of history anything that does not conform to its view of reality.

And it is—all too often, but not always —slavishly followed by *The New York Times*, which in deciding the facts "fit to print," has until recently totally ignored the evidence establishing the differences between the types of asbestos and the cigarette connection. Janet Chaplain of the Columbia School of Journalism established in a graduate school dissertation, that *The New York Times* has not reported any of the scientific studies on asbestos published since between 1976 and 1988. Yet the number of such studies is greater than the combined total of all those produced in the previous fifty years.

As a result the chasm between "the two worlds" of science and culture first described by the novelist and scientist, C.P. Snow, has never been greater:

"The world today is divided into two conceptual groups, the scientist and the nonscientist and the communications gap between them is wide and serious," Dr. Daniel E. Koshland, editor of *Science* wrote in the Oct. 25, 1985 issue. "What concerns me is that some of the fundamental concepts and methodologies of science are outside the understanding of the vast majority of the population, including its opinion makers."

As matters now stand, Koshland was only slightly exaggerating in his editorial in *Science* when he argued that "screening tests" in "scientific concepts" should be administered to "television anchors and gubernatorial candidates." Koshland continued: "Judges and legislators with little or no scientific training are making sweeping decisions on risk to the environment and from nuclear war and industrial accidents. Common sense would argue that an organization such as the Environmental Protection Agency (EPA) should list the major hazards to health and evaluate them systematically, taking the most important first rather than the most recent headline case.

"Scientists will be denounced for trying to introduce cold-blooded reason into an area where warm-blooded humanity is supposed to reign supreme. But warm emotion frequently gives way to hot-headed anger and even bigotry. The scientific method has been the most effective method of overcoming poverty, starvation and disease. Even those who are not professional scientists can understand its fundamental concepts, which will aid their decision-making in an increasingly difficult and technological world. It is time to bridge 'the concept gap' by improving scientific literacy."

That has still to occur, although the first fledgling efforts to bridge the gap were made back in the '70s when Selikoff's earlier critics had been effectively silenced. In 1978 he was testifying at an OSHA public hearing in Washington, D.C. asbestos would cause "40,000 excess deaths" a year due to asbestosis, lung cancer and mesothelioma. The use of such figures to justify a proposed OSHA policy—since abandoned—was to lead to an international scientific scandal, described in a later chapter.

But the scientific method was at work, quietly, unobtrusively, far away from politically dominated forums. Concern about asbestos emitted from a quarry in a nearby Washington, D.C. suburb of Montgomery County—the wealthiest in the country—was setting off a counter-reaction.

Dr. Malcolm Ross, a mineralogist at the U.S. Geological Survey, was prompted by the Montgomery quarry incident to survey the world literature on asbestos—not just what Selikoff had written. His study, published as a special technical publication of the American Society For Testing and Materials, concluded:

"The common 'white' asbestos has the least effect on those occupationally exposed, whereas 'blue'—crocidolite—has had the most effect. Despite the wide dissemination of 'white' asbestos—chrysotile— in our environment—in schools, homes, public buildings, brake lining emissions and so forth—there is little evidence that the very frequent nonoccupational exposure to this form of asbestos has caused any harm. On the other hand, nonoccupational exposure to blue— crocidolite—has been conclusively proven to have caused significant mortality."

Even before then, Doll in an article in the 1981 *Journal of the National Cancer Institute* report for Congress' own office of Technology Assessment and a book, *The Causes of Cancer* commented on the use that had been made of Selikoff's estimate of 40,000 "excess deaths" from asbestos as a justification for promulgating a massive carcinogen control program:

"The OSHA paper should not be regarded as a serious contribution to scientific thought," Doll wrote. "It seems that whoever wrote the OSHA paper did so for political rather than scientific purposes."

Weill was even more blunt: "It was a disgrace."

Nevertheless, the political debate over the relative health effects of asbestos, and the contribution made by cigarette smoking and the different types of asbestos was apparently over, finished as a matter of public policy-making in the United States. Asbestos was the villain, not cigarettes.

But nothing could stop other countries—with a far better record of dealing with the real dangers posed by asbestos—from continuing to do research and take action based on fact, nor fears. A 1985 British health commission, for example, after three years of study and hearings, directed by Doll, concluded there wasn't enough danger from asbestos to warrant its general removal from buildings. "Present evidence indicates that dangers from asbestos in buildings are likely to arise only when asbestos fibers are released into the air," the commission concluded, "or are damaged accidentally during maintenance and repair." The commission also concluded that the scientific evidence had established continued use of crocidolite or amosite could not be justified. Consequently, the British government prohibited their import from South Africa.

A Royal Commission in Ontario convened exclusively for the purpose studied asbestos exposure issues after $26 million was spent on removing the mineral from schools in the province. The commission was completely independent of any government agency, and went out of business after delivering a three-volume report in 1984. The commission members, who had no stake in building bureaucratic empires as do agencies like EPA, which serve as prosecutor, judge and jury in ban-hearings, concluded:

"Even acknowledging that the very young may be far more susceptible to asbestos disease, the health risks to children when asbestos fibers are released into the air remain insignificant because the level of exposure in asbestos-containing schools has in general been extremely low. The exception would be cases where asbestos material as actively disturbed or it fell on building surfaces and was disturbed.

"It follows with these exceptions, the program of removing asbestos from all asbestos-containing schools are not justified by the health risk posed to the students. . . . If anything, the scale and pace of the program significantly *increased* the risk to the risk to some workers directly engaged in control projects."

That was the message Dr. Robert N. Sawyer, who was emerging along with Ross as one of the two principal scientific critics of EPA regulations, was preaching. Sawyer, a medical doctor and engineer, had reluctantly come to have pronounced reservations about a man he had considered his "hero"–Selikoff. Sawyer, who had conducted the first asbestos removal project in the country, at the Yale School of Architecture, was becoming concerned about the extrapolation of Selikoff's figures from the workplace to schools.

Becoming the scientific advisor to the New York City Board of Education, he found many removal projects were being performed sloppily at best. By 1984, Sawyer, still then with Selikoff at Mt Sinai School of Medicine, was arguing at EPA hearings the cure was worse than the disease.

He argued EPA's policy was causing "more adverse health effects, including malignancies, than it is preventing:

"Currently there exists a common misconception," he said, "that the discovery of asbestos-bearing construction materials automatically indicates serious contamination and exposures in a school building.

"The surge in demand for removals greatly reduces the probability that a local school administration will obtain a competent contractor, knowledgeable architectural advice and a safe removal operation. There is a limited supply of such qualified personnel, and demand has already exceeded supply.

"The basic problem is that you're causing cancer in abatement workers by allowing this unnecessary removal," Sawyer concluded.

At the same time research into the health effects of the different types of asbestos was continuing, both in this country and abroad. Dr. John E. Craighead, a pathologist at the University of Vermont wrote in a *New England Journal of Medicine* article in 1984 that abdominal mesotheliomas occur only in persons exposed to the blue or reddish forms of asbestos. "Of all the types," Craighead said, "crocidolite [blue] asbestos is the most clearly associated with occurrence of the tumor."

Dr. Andrew Churg of the University of British Columbia followed up a 1985 study of chrysotile miners that proved any deaths of lung cancer were caused by trace elements of amphiboles, with an even more significant study in 1986. In that, he compared death rates among asbestos miners, people who lived in asbestos mining towns but never worked in the mines, and residents of a typical North American city, Vancouver. He found no difference in the death rates among the residents of the towns and Vancouver, even though the mining town residents had many times more chrysotile asbestos fibers in their autopsied lungs.

Ross cited a study of mortality patterns among women living in two Quebec mining towns. Their exposure to chrysotile had not proved to be "a significant health risk," he said. "If it were, the women of Asbestos and Thetford Mines, where over 20 million tons of crysotile have been mined, would be dying of asbestos-related disease. They are not."

Ross also noted that drinking water in the Thetford Mines area has "very high concentrations of chrysotile fibers, ranging from 172 million to 1.3 billion fibers per liter"—yet there was no evidence of increased cancer rates.

Three Vienna specialists reported in 1984 results of a comparable study in Austria from 1970-80. "No increased risk for lung or stomach cancer was found" in a mining town they wrote about in the *Archives of Environmental Health*: "No significant differences could be attributed to environmental asbestos exposure" near a plant processing the mineral.

Finally, and most conclusively, a 1986 study of Swedish 1,176 asbestos workers who were exposed solely to chrysotile, "did not indicate any asbestos related mortality," according to the authors, S.G. Ohlson and C. Hogstedt. The study, published in *The British Journal of Occupational Medicine* was definitive because 1) it dealt solely with chrysotile workers, and 2) using both national health insurance records and a national cancer registry that tracks every Swedish citizen from cradle to grave, it was able to follow up all but one percent of the cohort, a statistical completeness most researchers can only aspire to.

Selikoff's cohort studies were not nearly as valid as the Swedish study. His most closely contained cohort study was done on 922 men in an amosite factory in Patterson, N.J. But 270 were already dead when the study began, 39 had prior asbestos employment and, and 49 were lost to followup after termination of employment. In other words, well over one-third of the cohort could not be followed long enough to make any judgment whether they had experienced excessive mortality rates.

And real life experience was proving that asbestos-related disease was sharply declining rather than increasing. Asbestosis and mesothelioma are "exceedingly rare" diseases,' Dr. Edward A. Gaensler, director of a lung function laboratory at Boston City Hospital and a lecturer at Harvard told *The Boston Globe* as far back as 1983. He characterized asbestosis "as a vanishing disease."

Selikoff's own projections kept dropping over the years, from 40,000 "excess deaths" in 1978 to 20,000 at a 1980 press conference to 10,000 in an interview with the Associated Press on Sept. 287, 1881 to 8,300 in the 1983 *Globe* article. Finally, Selikoff, in a Sept. 29, 1986 *Newsweek* article conceded: "There's no question that a small amount of asbestos might be acceptable with good environmental controls."

The scientific method had worked. A theory had been advanced on basis of evidence which seemed plausible at the time, been reassessed by other scientists on the basis of new evidence and been rejected, even by its initial proponent.

But EPA had another agenda—and a myth of infallibility to maintain.

8 EPA: Washington's Magic Kingdom ▉▉

No other agency of the federal government has experienced less media or Congressional scrutiny than the Environmental Protection Agency (EPA). The agency might as well be the Wizard of Oz in his Emerald City, issuing dire warnings about the Wicked Witch of the West to a terrified populace.

But Dorothy learned the Wiz was just a small man with a big megaphone in a glass palace—and the most frightening thing about the Witch was her bargain-basement-bombazine dress. But Dorothy's companions weren't much help in her fight: Industry, the Tin Man who had no heart; the media, the Straw Man who had no brains; and Congress, the Lion who had no courage.

But Dorothy ultimately prevailed—with a cold bucket of water.

A modern-day Dorothy is Edith Efron, author of *The Apocalyptics: Cancer and the Big Lie,* which has been hailed by Dr. Bruce Ames, chairman of the department of bio-chemistry at the University of California at Berkeley, as "the *Silent Spring* of the counterrevolution." For Efron, more than any other writer or scientist, has thrown cold water on the Wiz of the environmental movement and EPA, and the excessive fear of the Witch of cancer both have spawned. But her bucket took years of meticulous study and research to make and fill.

However, even with her understanding of the way the environmental movement and EPA has manipulated science, the media and the public mind over the years, asbestos hysteria will continue to spread—and all the facts may as well be dammed.

"How Environmental Politics Controls What We Know About Cancer," was the subtitle of Efron's book. Those politics rest on four axioms, as described by Efron:

115

1. "Nature is a virtually cancer-free Garden of Eden rendered carcinogenic primarily by post-World War organic chemists," is the theory behind *Silent Spring*, according to Efron. That concept has never had any solid scientific support and has been rendered obsolete by the simple test developed by Ames, which has established that nature teems with carcinogens, including the male and female sex hormones, testosterone and progesterone, and the principal nutrient, lactose, in nursing mothers' milk.

2. "As much as 90 percent of cancer is 'environmental,' and by 'environmental'. . .anonymous informants at the National Cancer Institute meant the 'man-made' or industrial environment," Efron again pointed out. But Dr. John Higginson, first director of the International Agency for Research into Cancer, who initially made that observation back in the '50s, has made it repeatedly clear since then he gave greater emphasis to factors such as smoking, diet, sex habits, and cultural patterns than chemicals.

3. "Those who believed the post-World War II synthetic industrial theory of environmental cancer also believed, necessarily, that the cancer rates would explode," Efron wrote. In fact, with the single exception of smoking-related cancer, cancer rates have remained steady or declined despite an increasing larger—and increasingly aging—society. Indeed, nowhere has the decline in projected cancer rates been more dramatic than in those associated with asbestos, from an estimated 40,000-a-year to an actual 520-a-year in the mid-'70s.

4. The United States is "Number One" in cancer, "the carcinogenic sewer of the world," again according to Efron: "Presumably apolitical scientists attributed this to 'the Faustian sin' of being the world's superpower." But the Soviet Union pollutes worse than we do, as Susan Sontag observed in *Illness as Metaphor*. And statistics compiled by the American Cancer Society show the United States ranks last—25th—among industrialized countries in cancer rates. Luxembourg was number one with a rate of 219.14 per 100,000; England and Wales, number 10 at 191.19; Hong Kong, 16th at 173.67; and Australia, 21st at 161.48.

Efron then observed:

"I originally investigated these mysterious 'axioms' in order to assess the responsibility of the press in transmitting them," Efron wrote. "By the time I had finished, I had reached the conclusion that to blame the press under the circumstances. . .is absurd. The American press covers events, not ideas, and does not see past the policy makers. It [the media] has been nourished on bad science since the inception of the cancer prevention program.

"It was trained, as one trains a circus dog, to view apocalyptics, in and out of government, as fountainheads of scientific truth; it has shown a consummate credulity in the facade of arbitrary edicts brandished by the policy makers in the name of science; it has been taught by scientists to treat secretive or baseless assertions from scientific sources like political scoops.

"No one has taught the press that science does not operate by assertions, by leaks, by off-the-record briefings, by mimeograph machines spitting out documents with release dates geared to the evening news, or by documents with no names at all. No one has taught the press that the very appearance of such phenomenon means that what one is hearing is not science."

But all Efron's explanation does not change the fact that the media allowed themselves to be deliberately and systematically deceived by not only the environmental movement, but also EPA. The media, the Washington press particularly, allowed themselves to be used in a manner that would have long ago created an enormous uproar if the masters of deceit had worked for the FBI, the CIA, the State Department or the Defense Department. The media totally abdicated its responsibility to subject government agencies—not just those they distrust instinctively—to reasonably skeptical scrutiny.

"The deception lay in what we didn't say," as one public relations officer for EPA explained later.

"The main thing was that we tended to omit that we weren't able to do as much about the problems we were complaining about as we implied. . . . The idea was that our capability for retribution was better left to the imagination. Like a sheriff facing down a band of outlaws with one bullet left in his gun, we found it useful to pretend a fearsome posse was just about to round the bend behind us."

Those are the confessions of Jim Sibbison, a chief public information spokesman for EPA over the years, whose revealing article for *The Washington Monthly* of March, 1984 is titled, "Reporters: Prisoners of Gullibility." He reports that reporters who should have been presumably suspicious of any information emanating for an "Office of Public Awareness," which was the Orwellian title of EPA's public information office for many years, have been and are still being conned.

But that's largely because the national media wanted—and still wants to believe—in the Wiz and the Witch. The media, led at the time by Walter Cronkite, not only wanted to believe in EPA, but also its widespread coverage of Earth Day, April 22, 1970 created the impression that millions of Americans were up in arms about environmental degradation, according to a 1974 study by Ms. Peggy Wiehl, of the John F. Kennedy School of Government at Harvard University.

An illusion created a reality—and another illusion. When William Ruckelshaus accepted the position of administrator of EPA, an agency created by executive order of President Nixon, "the national news teams, including Walter Cronkite, came rolling to the door at 20th street where Ruckelshaus and company had made a home. Even though there was not even a legal agency yet, everyone expected full-blown competence, articulate EPA spokesmen, and the ability to do things as if they had been in operation for five years," Wiehl wrote.

"Ruckelshaus was forceful and dynamic, the image was one of momentum, of exciting things happening. The TV cameras focused on a little door sign that read 'Environmental Protection Agency;' the press felt the symbol was important, that the public should perceive that something new was in place."

But it was all—or mostly—images, symbols, perceptions, as Ruckelshaus candidly admitted later. The agency never knew what it was doing—scientifically—from the beginning. Ruckelshaus conceded readily to Wiehl: "The ability we had to review and analyze the standards at that time was almost non-existent. We inherited almost no analytic capacity. We just didn't have a very good assessment of what the options were, no analysis of what the real choices were as far as the statute was concerned, any secondary or tertiary effects."

So the agency went then—and now—the PR route.

"The key to Ruckelshaus' support," wrote Wiehl, whose paper was commissioned for a White House Office of Cabinet Affairs seminar, "thus was direct relationships with national environmental groups and with the national media. One problem faced by Ruckelshaus was the tendency of environmentalists to view public officials as either friends or enemies of the environment; one was either 'tough' or 'soft' on polluters. This polarity was reinforced by media coverage of environmental news. Environmentalists had easy access to the press and the ability to call the public tune on what was 'tough' and therefore 'good.' As Ruckelshaus put it, 'We couldn't even afford the appearance of being soft.'"

And so EPA, goaded by the environmental movement, lied consistently, methodically, and with clear and deliberate intention.

Sibbison explained how the P.R. process worked in practice:

"Our press releases were more or less true; the air and water really were dirty and we really were out to make them cleaner.... Few handouts, however, can be completely honest, and ours were no exception. The deception lay in what we didn't say."

Sibbison cited one major example, the Clean Water Act of the '70s. EPA's experience with that directly paralleled what was to happen with asbestos in the '80s:

"One major story we promoted in those pre-Reagan years was the discovery of cancer-causing chemicals in the nation's drinking water. Beginning in the early 1970s, the EPA had discovered that the chlorine use by city governments to disinfect the water was combining with otherwise harmless organic chemicals to form carcinogens. The leaders of EPA's water-supply program called for heavy publicity and we gave it to them."

With the single exception of the FBI under J. Edgar Hoover, and perhaps the original anti-poverty Office of Economic Opportunity (OEO), all federal agencies have been constrained by regulation, custom and the fear of provoking Congressional wrath from launching propaganda campaigns to promote its programs.

But that didn't stop EPA then or now. EPA was out "to whip the public into a frenzy about the environment" as Sibbison wrote, and political, if not practical, results followed:

"In 1974, a series of scare headlines about cancer and drinking water broke out in the headlines." Sibbison wrote, "Harold Schmeck of *The New York Times* caught the spirit when he covered a news conference conducted by Russell E. Train, then the EPA administrator. 'The Environmental Protection Agency,' Schmeck wrote in a page-one story for the *Times*,'" found drinking water of 79 American cities polluted with traces of organic chemicals, including some that are suspected of causing cancer. The agency has concluded that the problem exists throughout the country.'"

"By the time Schmeck's article appeared," Sibbison concluded, "the EPA had gotten what is wanted out of such publicity; a Safe Drinking Water Act signed by President Ford. Our efforts then turned to convincing the public the law would be effective.

Gladwyn Hill wrote another front-page piece for the *Times* when the new program went into effect The story proclaimed the dawn 'of a new era. . .in the supply of the nation's most widely used commodity, drinking water.' From then on, 'the purity of the drinking water and other aspects of the operation of 240,000 water systems was under the supervision of the federal government.'"

That was in 1977. There was a postscript, however, a follow-up story that the media has never printed.

"Since then the EPA [and, consequently, the media] has gradually come to talk less and less about drinking water," Sibbison wrote. "City governments have, alas, gradually caught on to the idea that our efforts weren't quite as awesome as Gladwin Hill said," Sibbison continued. "It turned out that the EPA wanted some big cities to install costly activated carbon equipment to filter out the carcinogens. The water treatment people refused, and that was the end of that. We have no reason to believe our drinking water is less carcinogenic now than it was ten years ago, but for obvious reasons, the EPA isn't eager to bring this up."

Sibbison went on to describe how the agency's "publicity bombardment" hyped concern over Love Canal into passage of the Superfund law. One result was a laudatory editorial in *The Washington Post:* 'EPA's decisions so far, although very slow in coming, appear to represent a model of a mature, cost-conscious but vigilant regulatory effort.'"

"We couldn't have put it better ourselves," Sibbison added.

"But how effective were our efforts in the 'following years," Sibbison asked. "I'm sure you don't know. [Since] 1981, we have seen very little in the press about this important program." Yet the news was there, if the mass media had been willing and able to go beyond the EPA propaganda.

One primary source of information was an article in the June 13, 1980 issue of *Science,* published by the American Association for the Advancement of Science. It contended the scientific study President Carter cited in order to justify the evacuation of Love Canal in Niagara Falls, N.Y. was "botched." *Science* went on to add: "Tragically, the EPA has needlessly terrified Love Canal residents."

Another source was a 1981 definitive study by the New York State Department of Health, which agreed there was "no evidence of higher cancer rates associated with residence near the Love Canal toxic waste site."

Even *The New York Times* conceded in a June, 1981 editorial, "From what is now known, Love Canal, perhaps the nation's most prominent symbol of chemical assaults on the environment has had no detectable effect on the incidence of cancer. When all the returns are in, years from now, it may well turn out that the public suffered less from the chemicals there than from the hysteria generated by flimsy research irresponsibly handled."

Eight years later, hundreds of Love Canal residents are being allowed to return to the homes they should never have been evicted from it in the first place. Most areas of Love Canal contain no more chemicals than other neighborhoods in the industrial city, David Axelrod, the state health commissioner, was quoted as saying in *The Wall Street Journal*, Oct. 3, 1988.

But Love Canal left behind a legacy that will plague the nation for years to come—Superfund. Superfund is a "largely ineffective, inefficient" colossus, the Office of Technology Assessment (OTA), a bipartisan consultation agency of Congress concluded in June, 1988. Superfund is plagued "by a lack of central oversight management and controls" that "causes inconsistency leading to confusion, unnecessary costs, and for some sites, ineffective cleanup."

The precedent set by Superfund has obvious implications if building owners across the country engage in an orgy of asbestos cleanups—and dumping, adding to the chaos already created at dump sites. "Lack of consistency among hundreds and eventually thousands of sites is not an academic issue," the report warned. "Harm to human health and the environment, loss of public confidence in government and wasting money are what's at stake."

This is nothing new, but rather part of a consistent and pervasive pattern that has dominated EPA attitudes and approaches over the years. Politics—and PR—have always dominated the decision-making process at the agency.

Ruckelshaus set the pattern back in June of 1972, four months before the next presidential election, when he banned the pesticide DDT, the primary target of *Silent Spring*. He did so even after his own administrative law judge, Edmund Sweeney, ruled following 17 months of hearings that DDT was not a cancer-causing agent. Indeed, the judge went so far as to declare: "Surely, it can't be seriously contended that the fact that DDT has NOT [emphasis in the original] been proven NOT to be carcinogenic is any logical reason for advocating a complete ban."

Ruckelshaus, nevertheless, went ahead with the ban for what he admittedly described were "'political,' albeit small 'p' reasons. Someone had to make the decision, and I was the person." Those considerations were obviously influenced by the political agenda of the man who had created the agency by executive order, Richard Nixon. And, although Nixon in his *Memoirs* denounced the DDT decision as "panicky," he was obviously happy to benefit from it in the national election.

Ruckelshaus relied on the Delaney Clause to the Food and Drug Act of 1958, which prohibits the use of carcinogens at any level in foods, as justification for the ban. The clause embodies the concept there can be no safe or threshold level of exposure to a carcinogen, a notion repudiated in the DDT hearings by most scientists, led by Dr. Jesse Steinfeld, the Surgeon General, to whose testimony Sweeney gave a "lot of weight" in his decision.

Subsequent events and scientific evidence have proven Steinfeld—and Sweeny—were right.

Frederick Coulston, former director of the Institute of Comparative and Human Toxicology and founding editor of *Toxicology and Applied Pharmacology*, has reviewed th results of the DDT ban in *CHEMTECH*, November , 1989:

"When some nations (e.g. India) banned DDT, disease such as malaria and filariasis reached epidemic proportions. Use of the chemical was resumed, with a consequent decrease in various insect-borne diseases. In addition, the safety of DDT to humans was apparent in these nations.

"No matter how much a person was exposed to the pesticide (it was even dusted on the skin to control lice), no ill effects were noticed. Other insecticides—notably the (commonly used substitute) organic phosphorus compounds were extremely dangerous. Furthermore, a high degree of skill is needed to mix these dangerous chemicals if serious harm or even death to the applier is to be avoided. Where that ability is in doubt, DDT can be counted as safe. From an economic point of view, DDT is attractive because, when used correctly for agricultural purposes, it is an excellent crop dust that allows more food to be grown safely and cheaply.

"In fact, few people realize that the use of DDT in the world today rivals the amount used during the time when the western countries also used DDT. It is still widely used in India, China, South

America, and Malaysia. In fact, WHO and the Food and Agriculture Organization (FAO) of the United Nations help developing countries to build factories to build DDT."

A joint committee of WHO and FAO experts addressed the three questions raised about the safety of DDT in 1984, and concluded, in Coulston's words:

"Is DDT a carcinogen in mice?

"DDT may cause liver nodules in mice, but these nodules never invade adjacent tissue nor do they metastasize, and therefore DDT is not a carcinogen in mice."

That conclusion is particularly significant because the scientific argument for the original ban was the fact that DDT did cause such liver nodules in mice, which could be expected to metastasize and become carcinogenic. The principal proponent of the argument at the hearings was Dr. Samuel Epstein, appearing as the principal scientific witness of the Environmental Defense Fund.

The principal drawback in the argument was dissected at great length in *Ecological Sanity*, by Dr. George Claus and Dr. Karen Bolander, a definitive history of the hearings: No such liver nodules were found in rats fed DDT. Therefore, if mice were not a good predictor for rats, they would be a less reliable predictor for humans.

And even rats aren't very reliable. Most rats fed carcinogens die well before the end of their normal life span. Ninety percent of all humans who die from cancer do so after the age of 70.

And, in fact, any evidence about any rodents and carcinogens is highly suspect. Rodents appear to be highly susceptible to cancer. Thirty percent die from cancer naturally. Further, testing with "maximum tolerable does" (MTD) may itself be carcinogenic."

Claus and Bolander described a study Epstein conducted in mice fed the equivalent of a human male drinking 25 gallons of gasoline for five consecutive days: "There is a growing body of evidence," according to Dr. Bruce Ames, discoverer of the Ames test for carcinogenesis, "that chemicals administered in animal tests at the MTD are causing cancer primarily by increasing cell proliferation, an essential impact of carcinogenesis. I think it is likely that a high percentage of all chemicals, both man-made and natural, will cause proliferation at the MTD and then be classified as rodents carcinogens, but this may only be relevant for high exposure."

Ames further argues if dosage levels of carcinogens were reduced, carcinogenicity would fall and so would the level of projected hazards because "humans and animals have numerous defense systems against mutagenic carcinogens which may make even low doses of mutagens protective in some circumstances."

Ames adds further: "Research is showing that low doses of carcinogens may be much more common throughout nature and much less hazardous than is generally thought." The main rule in toxicology is that the dose makes the poison.

Epstein, in his book, *The Politics of Cancer*, "sometimes described as 'the Bible'" of the environmental movement, justified maximum tolerable doses on the grounds that humans can be "less sensitive or more sensitive than rodents to the toxic or chemical effects of chemicals." For example, human beings are presumably 60 times more sensitive to thalidomide, which causes birth defects, than mice, while certain aromatic amines are "potent carcinogens for man, monkeys and dogs, but not for mice, rats and other rodents."

Therefore, he concludes "there is no known method for predicting a safe human level for carcinogens." Consequently, linear dose extrapolation is necessary, "which errs, if at all, on the side of safety for human beings." And he cited saccharin as the classic example, quoting Frank Rauscher, Jr., the director of the National Cancer Institute, as saying, "In protecting the public's health, there is no choice but to assume the extrapolation is linear."

Yet the no-safe dose concept should have lost all standing, politically as well as scientifically, after an attempt by the Food and Drug Administration (FDA) to ban saccharin was routed by a firestorm of letters to members of Congress in the late 70s. The National Academy of Science has subsequently recommended Congress modify Delaney to allow carcinogens be labelled with regard to potency into "high," "medium," and "low" categories, action Congress still lacks the courage to take. EPA, consequently, is still free to cite the "one fiber can kill hypothesis" in trying to ban asbestos.

Saccharin, historically, provides a classic example of the logical absurdity of extrapolating from high exposures to low ones— and assuming the risk to be the same. The first attempt to ban the artificial sweetener—so beloved of dieters—was tried by Dr. Harvey Washington Wiley, the founder of FDA back in 1908. Whiley's basis for arguing for the ban were tests performed on his "Poison Squad," young men fed increasingly large doses of chemicals as supplements to their diets.

Such experiments today would be considered not only unsound scientifically, but also criminally unethical, yet President Theodore Roosevelt, stout dieter that he was, didn't need those considerations to make up his mind. Roosevelt informed Wiley in no uncertain terms: "My doctor gives me saccharin every day. Anyone who says saccharin is dangerous is an idiot." Roosevelt convened a panel of scientists nominated by the principal university presidents of the time, which agreed with him.

Nevertheless, bureaucrats never let mere facts stand against unnecessary regulation. FDA Commissioner Donald Kennedy, for example, during the second controversy over saccharin 60 years later tried to use a study that presumably confirmed FDA's findings. That Canadian study suggested the normal use of artificial sweeteners increased the risk of bladder cancer in men by 60 percent:

But:

1. The study had never been peer reviewed by other scientists.

2. The Canadian study had never been published in the scientific literature.

Nevertheless, Kennedy declared it was sufficient evidence that cancer causation from saccharin was "virtually certain."

Since then, three additional studies, one performed by the National Cancer Institute—the largest on bladder cancer ever conducted—has found no increase in cancer associated with normal use of saccharin in either men or women.

The Delaney Clause has no more scientific acceptability than cows jumping over moons of green cheese. But it can provide a convenient crutch for bureaucrats and politicians to lean on. *Ecological Sanity* says Delaney was just such a political crutch in the DDT case. Claus and Bolander point out that if Ruckelshaus refused to ban DDT he "would have received very bad press coverage, particularly from the major East Coast dailies, and especially from *The New York Times*:

"[I]t would clearly have taken considerable courage and conviction for him to render a decision favoring the defenders of DDT."

The Kennedy School of Government report for the White House Cabinet seminar also independently confirms the political rationale behind Ruckelshaus' decision-making process: "The pressures exerted on EPA from the White House were greatly determined by President Nixon's personal view toward environmental protection and his calculations of political support," the Kennedy report read.

"While Nixon had no strong personal interest in environmental policy 'per se,' he saw the political need for at least the appearance of governmental response to pollution problems. The opposition of environmental groups was a potential thorn in his side particularly if Senator [Edmund] Muskie could marshall them in his Presidential bid."

Muskie, the front-runner for the Democratic nomination, was the leading proponent of environmental legislation in the Congress at the time.

Ruckelshaus was to display considerable courage before many months passed by being fired as Assistant Attorney General, along with Attorney General Elliott Richardson, for refusing to fire Archibald Cox, the special prosecutor looking into Watergate. But at the time the DDT decision was pending, he was still an ambitious young politician whose only previous experience with national politics was an unsuccessful run for the U.S. Senate from Indiana.

And as the first administrator of EPA, he was also fully aware of the opportunities such a post could offer. "A new agency like this comes along once every 15 years in the federal government," as he told Wiehl, the author of the Kennedy study, "and a chance to be in on the formation of such an agency was something you just don't get very often."

Even if EPA had possessed adequate knowledge and experience to rely upon to making a sound decision on DDT, political considerations may have tipped the scales. But, in fact, EPA did not have such scientific or analytic capacity.

In that absence—and without any effort to create that capacity—Ruckelshaus set two priorities for the agency, according to Wiehl:

"1) Generate public interest and support for EPA, and;

"2) put other polluters on notice that EPA was serious about enforcement.

"The idea was to get the media to help turn EPA into an enforcer that struck fear into the heart of polluters," Sibbison wrote in 1984. Ruckelshaus explained it candidly to *Time* magazine at the time: "The policy was to single out 'violators with the greatest visibility.' This, he said, 'got the message across' to those who might be tempted to follow the same path."

Sibbison explained further:

"The role of Ruckelshaus' press officers in this process was crucial. Call it hubris, but I came to feel that we press officers were the real journalists and the reporters were the publicists."

An "environment," for lack of a better word, had been created for the deliberate and systematic whipping up of public fears by the federal government, with the acquiescence of the New York-Washington media—without any basis in scientific facts established and accepted by competent, well qualified authorities in and outside EPA.

That atmosphere and policy climate was to change, but only slightly, during the passage of administrations. When the Reagan administration came in, EPA's press officers "found themselves marching to new orders," Sibbison wrote. "Previously, the EPA had sought to whip the public into a frenzy about the environment; now, a less, shall, we say, active administration wanted to try a cooler approach. Where once we had written freely in our news materials about the hazards of chemicals, employing words like 'cancer' and 'birth defects' to splash a little cold water in reporters' faces, suddenly we were urged to be more cautious."

EPA press officers, however, still felt free to regularly make up quotes for their bosses. By contrast, Larry Speakes, retired White House spokesman was forced to resign from a New York investment firm after admitting in an autobiographical book to making up one single quote for Ronald Reagan. The Speakes-Reagan quote, although in connection with world peace talks, was presumably unforgivable. Such journalistic sins are apparently much more easily ignored or even encouraged in a good–i.e. environmental–cause, as Benjamin Bradlee, editor of *The Washington Post*, and other top media executives have openly admitted.

One case cited by Sibbison was in metropolitan Los Angeles where a chemical dump approved by EPA exploded and burned.

"In the press release, I invented a quote for Anne Burford, as press officers frequently do for their bosses," Sibbison wrote. "(Sometimes the administrators get a chance to approve such quotes before they are sent out; more often, they never have any idea of what is being sent out on their behalf by their underlings.)," he added.

"'We acted immediately,' I had Burford saying, 'because the risk of another fire from about 5,000 drums on the site, I am told'" [wrote Sibbision for Burford–nice touch of verisimilitude, don't you think?] that the primary job now is to remove those drums and safely dispose of the remaining contaminated water.' [Burford's secretary] cut all this out too; apparently my ear was no longer up to EPA standards."

But Sibbison wasn't fired. Nor did his later confession in *The Washington Monthly* receive any attention from the Washington press corps, a classic example of media-hypocrisy.

But that's not to say Reagan conservatives weren't willing to play fast and loose with scientific facts–and people's lives–when it suited their purposes. Politics was apparently the primary consideration in the Burford decision to evacuate the citizens of the small town of Times Beach, Missouri, because of possible exposure to dioxin, the contaminant which may make Agent Orange a carcinogen. The decision to order the evacuation February 22, 1982 came at a critical time for the agency, according to *Chemical and Engineering News (C&EN)* of June 6, 1983.

"EPA Administrator Anne Gorsuch Burford was under intense pressure from Congress for information related to the actions the agency had or had not taken under the Superfund law, and the body was moving in early December toward finding Burford in contempt of Congress.... Then, in early February, the EPA administrator in charge of waste cleanup and Superfund, Rita Lavelle, was fired by President Reagan, amid allegations she had used Superfund for political leverage.

"Possibly pushed a little faster than she would be otherwise, EPA Administrator Burford announced February 22 that the federal government would buy up all the contaminated property in Times Beach, paying the residents at prices reflecting property values before the contaminant dioxin was found," *C&EN* reported.

However, a process that was to have taken no more than 90 days was still snarled in red tape more than two years later, according to a 1986 *Detroit News* series. And the people of Times Beach had been stigmatized and avoided as personally contaminated by their neighbors. The psychological damage was probably far greater than that caused by exposure to dioxin at levels lower than those found in many industrial communities. Dioxin is a by-product of many high temperature processes, including incineration.

The Times Beach evacuation, unlike Love Canal, was vigorously protested by the scientific community in the immediate aftermath. But the protests went virtually unnoticed by the mass media except for a brief notice in the *Washington Post* of June 23, 1983, which reported:

"The 351 members of the House of Delegates of the American Medical Association (AMA) voted to adopt an active public information program. . .to prevent irrational reaction and unjustified public fright and to prevent the dissemination of possible erroneous information about dioxin."

The resolution introducing the proposal noted only two short-term reactions have been observed from exposure to extremely high levels of dioxin, primarily a form of acne called chloracne. Both clear up within weeks or months after removal from exposure.

"The news media have made dioxin the focus of a witch hunt," the resolution continued, "by disseminating rumors, hearsay and unconfirmed, unscientific reports. . .attributed to scientists whose quotes should have been, 'I don't know.'"

Even Dr. Irving Selikoff, as the specialist in cancer presumably caused by asbestos cited most often by EPA, agreed with the general thrust of the AMA regulation, although being the source of many of the confirmed and unconfirmed reports on environmental cancer quoted by the media. This time, he was quoted in the *Post* article as agreeing, at least in part, with the AMA resolution:

"We should not be saying things we don't know about. . . . In the short term accidents it is true we do not see disease." Selikoff, however, then weaseled out with the caveat usually cited by scientists associated with the environmental movement: "We don't know its long-term effects"—in other words the impossibility of proving a negative or that dragons don't exist.

The unwillingness or inability of the mass media to consistently and systematically follow up the known scientific facts behind stories on issues such as Love Canal, the Clean Water Act, Superfund—not to mention nitrites, cyclamates, saccharin, Agent Orange—is not only needlessly wasting billions of dollars but literally killing people by emphasizing sensational nuisances rather than quiet—but real—killers.

Dioxin provided a classic example of the media's convenient ignoring of inconvenient facts when Dr. Vernon Houck of the Center for Disease Control admitted in the Spring of 1991 that Times Beach, Missouri, should never have been evacuated. Houck was personally responsible for recommending the evacuation. The amazing turnabout, which has prompted a $6 million restudy of dioxin by EPA, was virtually ignored by the media until an op-ed piece by Reed Irvine of Accuracy In Media appeared in *The Wall Street Journal* August 6, 1991. Nine days later on August 15, *The New York Times* ran its first complete report on dioxin, an account that started on Page 1 and took up several columns inside.

In retrospect, Dr. Houck said, "We should have been more upfront with Times Beach people and told them, 'We're doing our best with estimates of risk but we may be wrong.' I think we never added, 'but we might be wrong.'"

Facts were known and published years ago, which, had they been followed up, would have prevented the present asbestos situation from ever occurring. OSHA had tried to develop a national or "generic" standard for identification of carcinogens, which could be used by all federal agencies. The OSHA attempt was to be quietly abandoned several years later, after the attempt produced an international scientific scandal of enormous proportions—linked directly to asbestos.

The proposal for the standard was developed by a private company with previous links to a top OSHA official without consultation with the National Institute of Occupational Safety and Health (NIOSH), the agency established by Congress to provide OSHA with scientific and technical advice. "NIOSH was completely bypassed," Edward Baier, the institute's deputy director at the time, said, "To what purpose, to what end I cannot say."

One reason was probably the fact that NIOSH officials testified that only three to five percent of cancer was caused by industrial exposure. Consequently, there was little or no reason for the proposed standard. Shortly thereafter, a mysterious paper which soon became notorious in the scientific community as "the estimates document" was issued by the Department of Health, Education and Welfare (HEW), now the Department of Health and Human Resources (HHS), appeared. The document took projections of cancer cases caused by asbestos developed by Selikoff and others—particularly cases associated with cigarette smoking—and upped the percentage from three-to-five to 20-to-40 percent.

The names of eight—later nine—prominent scientists appeared on the cover sheet of the document as "contributors." The names were alphabetized, meaning that no individual in the group took responsibility for the data. "Who then had written the study?" Efron asked rhetorically. The answer was obvious: 'Nobody. . . .'" It was an immaculate a.k.a. bureaucratic product.

Nevertheless, *Science* took dead aim on the estimates: "These estimates are clearly inflated. In each case, the investigators have taken the highest risk ratios available—ratings obtained from workers exposed to massive concentrations of the carcinogens—and multiplied that by the total number of workers who might have been exposed to the carcinogen, even though most of the workers had never been exposed to the concentrations on which the risk ratios were based. In a similar analogy, one might find that risk of a driver dying in an automobile crash is one in ten if the automobile is driven consistently in excess of 120 miles per hour, and there are currently 100 million automobiles on American highways. Using the logic of the government report, one would conclude that there will be 10 million excess deaths as a result of driving at high speeds."

Science also took a hard look at the figures on shipyard workers studied by Selikoff, and still being used by EPA to justify its proposed ban on asbestos. "Some four million of these shipyard workers were exposed to asbestos in World War II. The HEW investigators predict about 1.6 million are expected to die of asbestos related causes—enough to account for 13-15 percent of all cancer deaths in the United States.

"Yet only 1,000 cases of mesothelioma are reported annually—not the 7,000 cases that would be expected if the prediction is to become true. And, at that rate, every worker should be dead by now. Instead, 40 percent are still alive.

"Severe deficiencies [exist] in the original report and necessitate the conclusion that its predictions are invalid," the article concluded.

The *Science* article was not alone. The "estimates document" was also ridiculed in *The Lancet*, the official publication of the British Medical Society, and the subject of a severely critical Congressional report, which went so far as recommend the disbanding of OSHA.

The report, written by Richard Zeckhauser, professor of political economy, and Albert Nicholas, assistant professor of public policy at the John F. Kennedy School for the Senate Committee on Governmental Affairs, concluded: "Rather than continue on the course of its first seven years, we would argue OSHA should be disbanded. Safety and health in the workplace would not suffer measurably, significant private and governmental resources would be saved, and an agency seen primarily as a source of government harassment would be eliminated. . . .

"Unfortunately, the current proposal for generic standards reveals little, if any, concern with improved decision making. The sole objective appears to speed up the process of issuing standards..

"OSHA has learned in the safety area, to its sorrow, of the dangers of overly hasty action. Now is not the time for OSHA to lock itself into a rigid framework for making regulatory decisions in the field of occupational health. The risks of precipitous action are simply too great, not just in terms of cost, but also in terms of workers' health, since the unfocused approach of the current proposal will dissipate scant resources, including OSHA's, leading inevitably to the neglect of truly serious hazards."

The words of Zeckhauser and Nicholas are variations on a theme sounded many times since then—and before. Gov. James Martin of North Carolina, for example, spoke before the original OSHA hearing on the generic carcinogen policy. Martin was not only a member of Congress at the time but also the only one with a doctoral degree in chemistry. He first cited the National Academy of Science study recommending revision of the Delaney Clause for carcinogens to be classified with regard to potency into "high," "medium," and "low" categories, as a response to the FDA attempt to ban saccharin, and then observed:

If OSHA sets out "to ban or economically cripple products [asbestos] whose risk is comparable to saccharin, you will not thereby reduce the statistical risk of cancer, you will divert and dilute attention that should be focused on serious carcinogens, you will confound and confuse the public and you will deserve the ridicule you will receive.

"That would be a tragic parody of the responsibility you ought to exercise."

Former Sen. Richard S. Schweicker, as secretary of HHS, in a letter dated April 29, 1982, effectively repudiated the "estimates document:" "More recent estimates imply that original estimates of cancer associated with asbestos were too high. Current estimates for overall workplace associated cancer mortality vary within a range office to fifteen percent. Recent estimates of the contribution of asbestos to cancer mortality range from one to three percent

"The central issue," Schweiker concluded, "is that all of us must do what we can to make the workplace less hazardous for those who continue to work with materials like asbestos for which society has not yet found a substitute."

These reports, studies, records and letters are matters of public record, but have gone almost completely unreported in the mass media.

Why?

Sibbison has one answer—and some concern about the consequences. "[T]he laziness of the press was initially helpful to the cause of cleaning up the environment because it led reporters to make the EPA out to be a ruthless enforcer of its policies. In the long run, however, the willingness of the press to allow EPA's press officers to manage the news proved harmful. . . . [M]ost reporters have made little effort to learn where the EPA falls short of its promises. The result has been that public allows the EPA's failures to continue."

That failure will continue if the fast-growing asbestos abatement industry obeys the exhortations of Ruckelshaus at the annual meeting of the National Asbestos Council (NAC) in April, 1988. Ruckelshaus first acknowledged environmental concerns were largely a media phenomenon, first heightened by the introduction of color television in the late '60s and early '70s.

"Believe me," he said, "the problem of pollution depicted in black and white has nowhere the impact it has in color." Then he urged the NAC membership to get involved in the regulatory process. "It's important for you to develop the capacity to bring to the public policy debate your knowledge, your understanding and your hands-on experience of what these laws, rules and regulations should provide."

That's a little like having the pigs offer opinions on whether the troughs should be made of mahogany or cherry and covered in satin or chintz.

Soon after delivering the speech, Ruckelshaus became president of one of the largest asbestos abatement companies in the country.

"It was beautiful and simple as all great swindles are."

–O. Henry

9 Vigilante Moms

"Getting the mothers to form a vigilante mob and storm the school committee" was the theory behind EPA's 1982 regulation requiring public and private elementary and secondary schools to inspect for the presence of asbestos.

The rule had nothing to do with science or public health. EPA's more than 100-page rationale for issuing the regulation was torn apart by outside scientists the agency itself retained to review the so-called scientific justification. Other EPA studies showed at the time removal was killing abatement workers while providing negligible health benefits for building occupants.

Still other studies showed removal operations increased ambient air levels in the atmosphere far higher than they were before the projects began. The regulation, was in fact, just a ploy, a gambit to shift responsibility for asbestos control from EPA to local officials.

William Ruckelshaus was not the administrator when the rule was promulgated. But in a lengthy interview with me he showed no signs of disagreeing with the rule or the intent behind it:

Q. "Why are you requiring the schools to inspect for asbestos and not other kinds of buildings?"

A. "It will move to other buildings and commercial buildings as well. It was the schools where it started."

Q. "What was the real reason behind the school rule?"

A. "It was simply a question of who was going to pay for it, who was going to pay to clean it up. . . . The approach that was adopted by the Administration before I got here was the burden was to be borne by the people living in the school districts. . . That's what the fight is all about. Who's going to pay for it."

Q. "An EPA lawyer told me the theory behind all this is to whip the mothers up into a vigilante mob to storm the school committee to do something."

A. "You could put it that way. Obviously, the purpose of the approach was to bring pressure on the school boards to correct whatever situation they found.

"First of all, do the inspection. Do you have a problem? If you don't, don't worry about it. If you've done the inspection and find you have a problem that ought to be addressed, then you should notify the parents. If you notify the parents, the parents will bring pressure upon you to correct it."

Q. "The law requires schools to inspect for the presence of asbestos. Why not write something into it that would have required some action on their part?"

A. "Well, you could."

Q. "Well, why wasn't that done?"

A. "The agency could do it. . . . There's the authority in now for the agency to order action. The theory was that if the federal government ordered the action, what would rapidly follow was a requirement that the federal government pay for whatever action was ordered.

"That was what scared them.

"And if the federal government had to pay for it, you could just bet that the cost would be several orders of magnitude greater than if you had some balance with whatever risk reduction you wanted to achieve.

"If you removed the requirement of paying from the people who are going to benefit, which is exactly what we've done with these hazardous waste sites [Superfund], the demand then is for zero risk. And the cost just goes straight up, because it's a cost free benefit for them. Why shouldn't they demand zero-risk?

"And so, the theory is you've got to keep the cost and the benefit within the same person. And then you can force them to come to grips with what sort of reductions make sense, what is reasonable to do under the circumstances.

"Now that was the theory behind trying to force the school

boards to do the inspection, notify the parents and then the parents would come in and make that kind of balance."

Q: "So that was a deliberate decision?"

A: "Absolutely."

Q: "So, what you're doing is essentially putting cost-benefit analysis on the agenda of every community in the country that supports a school system?"

A: "Sure, absolutely. We try to present them what we think the risk is, what can be done to correct it—force them to go and find out if they have any risk and then force them to make a judgment as to what you should do about it.

"Now the question of whether you panic them or whether the whole thing is going to do more harm than good, frankly I think that's there, any way you do it. I don't care whether the federal government does it or the local government does it, I think you've got the same problem.

"But putting the onus on the person to make the judgment as to how much risk they want to reduce as long as the same guy is going to pay for it, it's in my view sensible. You take that away from them, which is exactly what we've done in the Love Canal kind of situation, they just demand zero risk—they want that stuff out of there. 'Don't tell me how much it costs. We don't care.' And why should they?"

Q. "Dr. Bob Sawyer of Mt. Sinai Hospital [now at the University of Pennsylvania] has said flatly that EPA's asbestos policy is going to kill people."

A. "Well, you know, whenever we have a problem like we have with this asbestos all over the country. . .and we have no governmental apparatus to grapple with it, the potential for people panicking. . .is there. Any program that we would have had—whether the government moved into it, with a training program, with a contractor-certification program—still would have been faced with some people deciding, once they had discovered that they had asbestos in their buildings or their schools, to go forward without waiting for the government to take hold. That potential is there.

"Our people were terribly worried at EPA about whether we

would have people coming in and trying to remove asbestos, and potentially causing a worse problem than was there before. That potential is there—there's no question about it.

"What we have consistently recommended is that they be very careful about how they proceed, but they be careful about how they do it. And we provided them guidance. And the question of whether or not that they could go ahead and create a worse situation, the potential is there. There is no doubt."

Q. "Malcolm Ross of the U.S. Geological Survey is saying not just that there's panic, but there's a very good reason to argue the panic is justified."

A. "I've listened to him have a lengthy discussion with our scientists. I sat in a room for a couple of hours and listened to them discuss this whole issue of blue and white asbestos, and which one was really carcinogenic and which one was not [carcinogenic]. Our people believe, our scientists believe very strongly that while one—I can't remember which one—is far more carcinogenic than the other, they're both carcinogenic.

"Those kinds of uncertainties—whether its EDB, DDT or asbestos—are all part of this equation. There is enormous scientific uncertainty is this field. . . . That doesn't mean to say this guy [Ross] is wrong, and these other scientists are right. I mean there isn't any right or wrong here. It's just a question of where the consensus came out. The consensus came not supporting the position as he was espousing it within the government. He may prove to be right.

"It's one of the most frustrating parts of the job of being administrator of EPA. It is when you're operating in these areas of enormous scientific uncertainty, even though you try to make it as clear as you can that this uncertainty exists, people continue to take a number out of a risk assessment methodology and write it in stone, as if that number were exactly how many deaths were going to occur over a particular period of time. And we don't say that.

"We indicate that this is the upper bound of the risk; the lower bound can be all the way down to zero. But based on the risk as we know it and what can be done about it, here is our judgement on how we ought to proceed.

Q. "Is there a rational standard of asbestos exposure?"

A: "Our scientists have consistently told me we don't know how to set an ambient air quality standard for asbestos that makes sense. We don't know how to do it.

"So, what we have suggested instead are these guidelines or approaches where you find problems of a certain magnitude in a school, commercial building, wherever you find it. And the guidelines, in effect, take the place of the standards.

"With asbestos there is always some residual risk. . .where a fiber can provide a risk is someone lives long enough."

Q: "Do you think environmentalists exist not to be satisfied?"

A: "That's true, but those people don't necessarily dominate the bureaucracy in EPA. One of the problems in this whole field. . .[is] the paranoid style of American politics which so dominates the thinking. 'The Congress writes these laws as though the executive branch is not to be trusted in implementing them. . . . They give us deadlines in which to act, which in most cases are unreasonable. They set standards for achievement, usually zero risk, which we can't achieve either.'

"And then if we don't do that, we get a citizen's right to sue provision and turn the whole thing over to the courts. I think that's a terrible way to proceed.

"What we've done is divide this responsibility between the Congress and the executive branch and the courts, and, in effect, locate the power to act nowhere. And we just bounce back and forth between these branches of government—without in my view, providing the clear criteria for balancing risks and benefits and costs that the agency ought to be balancing.

"Instead, all this responsibility is mixed between the Congress and the executive branch. Getting the Congress to respond differently as we have traditionally in dealing with these problems is something that just defies me."

Much of Ruckelshaus' criticism of Congress is certainly justified. But he was less than candid during our interview in failing to note it was Congress, specifically the House subcommittee on appropriations for EPA chaired by Rep. Edward P. Boland (D. - Mass.), which forced the agency to revise language in its guidebook for schools which encouraged removal rather than maintenance in place. "Removal is the only permanent solution," were the chilling words expunged from the revised *Guidance for Controlling Friable Asbestos-Containing Materials in Buildings* —but only after express action of Congress over the pained protests of EPA. Further, nowhere in the 200 page guidebook was any mention made of the differences between the types of asbestos. Again, other than in the interview with Ruckelshaus, EPA has never admitted to the "enormous scientific uncertainty," never mind the "paranoid style of American politics" that has characterized EPA's rulemaking process. And the agency has never told local school officials the inspection process is designed to stimulate cost-benefit analysis, rather than panicky removal projects.

Instead, EPA has deliberately turned mothers into vigilante mobs. The agency has incited the panic it publicly deplores. "The behavior of the EPA on this entire issue is absurd," John J. Sweeney, president of the Services Employees International Union (SEIU), whose 85,000 members include many school custodians and maintenance workers, has complained:

"The ultimate absurdity is that EPA has created a high level of concern about asbestos, but refuses to tell anyone how to deal safely with the substance."

"EPA puts the weight of the burden on individuals to encourage the removal of asbestos from the environment of children," William Borwegen, safety and health director of SEIU testified at a hearing convened after the union sued the agency in an unsuccessful 1985 effort have formal standards for dealing with asbestos set.

"They potentially set friend against friend, teacher against administrator, parent against school and children against all of the above." And Ruckelshaus made his own contribution to this civil war by resorting again to PR techniques to heighten tensions—rather than communicate knowledge.

On September 28, 1983, he sent a memorandum to EPA

regional administrators entitled "Communications Strategy for Asbestos in Schools Program." In it, he supported the "use of press releases" to secure compliance with the regulation, but actually to force school officials to consult parents and school employees—" getting the mothers to form a vigilante mob to storm the school committee. A sample release was included, with useful instructions for achieving maximum exposure, such as: "Case by case releases may quickly exhaust the public's attention to the issue. The use of releases which cover a 'batch' of inspections, which report the compliance status in a single geographic area (e.g. the Bay area), or which announce regional trends should avoid overplaying what we remember is a very emotional issue."

Despite the warning, the release contained such inciting language as: "Asbestos fibers can be released to the air. They are harmful when inhaled, and can cause asbestosis, a lung ailment, or cancer of the lung and the gastrointestinal tract."

Further language was included which might well have the effect of crying fire in a crowded theater, which Justice Oliver Wendell Holmes cited as an example of the kind of free speech which does not enjoy First Amendment protection. That should presumably be as true for government agencies as individuals. But, apparently, EPA considers itself exempt from such restrictions.

"While the EPA rule does not require the removal of asbestos-containing materials, school administrators are encouraged to discuss this problem with parents, custodians, teachers and qualified professionals to determine if corrective action is warranted," the guide stated. But, of course, the agency has never provided school administrators with any objective information permitting a rational assessment of the problem.

And EPA has never put out any press releases announcing the fact that its own criteria document justifying the issuance of the original 1982 regulation was savaged by outside experts retained by EPA itself to assess its merits. The four independent reviewers picked apart the EPA's analysis of asbestos-risk data in buildings as "unconvincing," "greatly overestimated," "superficial," "scientifically unappealing," "judgmental," and "absurd." One scientist on the panel said the agency's conclusions reflected "pre-conceived notions" and an "underlying tone of bias."

EPA solicited comments from the four professionals as part of its normal rule-making procedure. But the critics' harsh language aimed at a 1980 draft of a "scientific support document" to build a foundation justifying the school-inspection mandate was ignored.

The final support document was issued without any of the major changes warranted by the critic's reviews, in January, 1982. The actual inspection rule was issued by EPA four months later.

The 138-page support document was produced mainly by two agency employees—Harry Tietlebaum, who had, at the time, a doctorate in biochemistry, and Charles Poole, who holds a master's degree in public health.

They adapted as their primary source material, indoor asbestos-exposure data developed by Dr. Patrick Sebastien, now at McGill University in Montreal, for the French government. They also cited earlier cancer-rate studies by Dr. Irving Selikoff, now retired from Mt. Sinai Hospital in New York.

Poole and Tietlebaum's preliminary study concluded "building materials in common use can release asbestos fibers" and expose occupants to increased risk of "numerous serious and irreversible diseases." The school-inspection rule, released four months later, used similar language and spoke of a "significant hazard to public health."

The two government scientists, Teitlebaum and Poole, concluded that children and adults in schools across the country might be exposed to an average of 57 nanograms (a microscopic weight) of airborne asbestos each day if they spent five to eight hours time in various parts of a building containing asbestos. If they stayed in areas of heavy activity, gymnasiums and corridors, for example, the maximum daily exposure could rise to 270 nanograms, the EPA report estimated.

As the buildings deteriorated, from use and age, the maximum exposure could be anticipated to rise to 500 nanograms, the report said.

A nanogram is one-billionth of a gram—a value comparable to three heartbeats in a 70-year lifetime, Ross, of the Geological Survey, has estimated. Asbestos fibers are extremely light, and the EPA calculated that 30 would weigh a nanogram. The ambient air in St. Louis, Mo., a typical American city, contains between 17 to 32 asbestos fibers, official guidance samples taken by the National Institute of Standards and Technology of Standards have established. The samples are used in comparison measurement, under electronic microscopes to determine whether the asbestos levels in buildings are too high.

The EPA federal support study said its estimate assumed the exposed population is distributed over all U.S. schools with friable asbestos-containing materials, and "these areas have the same distribution as the measurements of Sebastien.."

But those assumptions and the overall conclusions about asbestos dangers cited in the report were criticized sharply by the outside scientists hired to provide "peer review" of EPA's. Such peer reviews are a standard government procedure for assessing the quality of scientific research.

One of the four outside experts was Julian Peto of Oxford University in England. He not only analyzed the use of Sebastien's figures, but directly contacted the researcher in Paris. Peto's report back to EPA was scathing.

"The assertion that Sebastien's data are applicable to U.S. schools is unconvincing," wrote Peto, who has a doctorate in mathematics and is an internationally known medical statistician. "Dr. Sebastien, whose measurements in Paris buildings are claimed to be representative of U.S. schools, does not regard this assumption as valid."

At another point, Peto added: "This maximum estimate of average exposure (270 nanograms) is absurd."

Even EPA's calculation of how much asbestos fibers weigh was challenged. "I could not persuade him [Sebastien] to offer an estimate of the appropriate conversion factor from nanograms to fibers, assumed to be 30 in the present draft," related Peto, the Oxford professor.

Frank Carlborg, another one of four who has a doctorate in statistics and taught at two Illinois universities, concluded in his peer review: EPA's methods and conclusions were "scientifically unappealing."

Yet another commentator paid by EPA, Dr. Philip Cole, told the agency: "The selected estimate of prevalent asbestos exposure probably greatly overestimated the true levels." Cole, an epidemiologist who taught at Harvard and now is at the University of Alabama, alleged that the draft of the 1982 document contained an "underlying tone of preconceived notions of the risks associated with airborne asbestos in schools.

"For example, the first paragraph of the introduction seems to conclude that the occupants of the schools are incurring a risk of developing asbestos-related disease. This is completely judgmental."

Cole also disputed the widely cited studies of Selikoff and two colleagues cited by Teitlebaum and Poole. Selikoff et al concluded in 1964 that lung cancer rates among 1,522 asbestos insulation workers were "found to be at least seven times as common as expected and cancer of the gastrointestinal tract three times as common as predicted."

Selikoff and his team linked asbestos and cancer after studying three groups:

● 622 unionized insulation workers in New York and New Jersey. Their survival rates were monitored from 1943 through 1976, as were the causes of death.

● All 17,800 members of the same union in 120 U.S. and Canadian locals. The study recorded their death rate, causes and years since first asbestos exposure.

● 933 workers in a Patterson, N.J. factory that made insulation from reddish brown (amosite) asbestos. Their health was tracked from 1943 on.

Based on data from those workers Selikoff concluded: "asbestos insulation workers in the United States and Canada suffer an extraordinary increased risk of death by cancer and asbestosis associated with their employment. . . ."

But Selikoff's research data was seriously, if not fatally flawed. He compared all the asbestos workers only to the general male population, not to control groups drawn from the general work force, particularly blue collar workers and especially those exposed to high concentrations of dusts. Moreover, Selikoff's own work noted that many of the subjects studied handled asbestos "in rather tight quarters," a situation likely to have increased their health risks.

Further even Selikoff's studies, as well as others, particularly ones done abroad, have suggested for years that disease patterns were—and are—influenced strongly by personal smoking habits as well as the kind of asbestos involved.

Accordingly, Cole warned EPA against relying heavily on Selikoff's broadly generalized studies linking asbestos to elevated cancer rates:

"An unwarranted degree of accuracy is attributed to these estimates," Cole wrote in reviewing the agency's support document. "It should be appreciated these estimates are subject to considerable error.

"Applying the risk estimates obtained from the insulation workers to school occupants [creates] the potential for gross confounding by cigarette smoking. Approximately 80 percent of the insulation workers were either current or ex-cigarette smokers."

Cole also noted that the Selikoff team's own data "indicates that approximately 88 percent of the lung cancer deaths of asbestos workers could have been prevented if none of the asbestos workers had smoked."

The fourth reviewer of EPA's support document was Dr. Kenneth Rothman, who has a doctorate in epidemiology and biostatistics: "The main objection to the document as a whole," he wrote, "is the superficial way by which the actual risk estimates were derived. . . ."

Rothman echoed Cole's concern about the industrial data gathered by Selikoff.

"The final calculations are highly sensitive to the unstated assumption that the excess deaths among the insulation workers who died are indicative of the experience school children will face each year of a life for a lifetime."

However, despite the unanimous criticism of EPA's own experts hired specifically to review its justification for the asbestos-in-schools rule, the agency altered none of its key judgments in the final 1982 regulation. The final rule said:

"Exposure to asbestos fibers can lead to numerous serious and irreversible diseases. . . . In particular, friable asbestos-containing materials have been found to release fibers in concentrations sufficient to increase the risk of these diseases for the building occupants. . . . The agency has determined that exposure to asbestos in school buildings poses a significant hazard to public health."

Despite underpublicized disclaimers, EPA's continuing case against asbestos rests on two premises:

● The presence of any asbestos in a building is dangerous.

● Potential risks can be projected from medical studies of workers exposed to high concentrations 20 to 40 years ago under completely uncontrolled conditions.

The premises thereby adopt the thesis or belief that first, "one fiber can kill." Second, any and "all asbestos exposures, even those of very brief duration or very low intensity, increase the risk of cancer," as the scientifically repudiated support document states.

But even Ruckelshaus admits the premises are confounded by the facts.

"People continue to take a number out of a risk assessment methodology," Ruckelshaus complained, " and write it in stone. . . . We [EPA] didn't say that."

But EPA certainly created that impression, despite Ruckelshaus' disclaimer. Further, EPA has shown a remarkable capacity to avoid being confused by the facts—even its own facts.

A year after the asbestos-in-schools rule was issued, the agency issued a booklet which admitted it's own original rule was impossible to prove one way or the other. "EPA and the scientific community believe that any level of asbestos involves some risk, although the exact degree of risk cannot be estimated."

Aside from EPA's gratuitous assumption that it speaks for the scientific community, the agency was really just restating the same old environmental movement claim: We may not be able to prove asbestos does cause cancer; but can you prove it doesn't?

That obviously is being asked to prove a negative, a logical as well as a scientific impossibility.

But other arguments do abound, taken from reality. The results of EPA's own asbestos policies and actions are the agency's most devastating critics and the best evidence of their own futility.

For example, one EPA study estimates, on the basis of actual surveys conducted by its regional asbestos officers that 75% of all asbestos removal projects were improperly conducted.

Another study, a cost-benefit analysis run by EPA itself–this one calculated not in dollars but in terms of human lives–came to light through a suit filed by the SEIU.

The study, which was completed by the summer of 1984 by the agency's Clean Air Division, estimated 4,414 cancer cases would almost certainly be caused by sloppy removal operations. Such sloppy operations were previously estimated at 75% of the total of such projects.

Conversely, maintenance programs, which had been spurned by the agency until Congressman Boland's appropriations committee intervened, would have prevented 104 cancer cases.

No press releases were issued to announce the existence of this study. Indeed, the title page was emblazoned with the words, "Do Not Cite or Distribute."

The reason given for effectively suppressing the report by Michael Shapiro, an EPA spokesman, was that it lacked peer review. That logic obviously fails to provide any justification for not publicly announcing the devastating peer review of the original justification document.

The intellectual dishonesty behind this situation was addressed in a *Detroit News* editorial of August 1, 1985, if one may be permitted a bit of self-congratulation:

"Michael Bennett, the *Detroit News* special reporter who first poked holes in the EPA policy and the weak science that supported it discovered the study after the EPA had quietly changed its instructions to say that the presence of asbestos was not necessarily hazardous to the occupants.

"Congress did not know of this report until a few weeks ago. Even so, the House Appropriations Committee withheld $50 million in remedial grants for poor school districts, until the EPA agreed to publish a new guidebook for school boards that says the presence of asbestos in buildings does not necessarily endanger occupants, and that a crash program for removal can do more harm that leaving the stuff in place.

"As we said, the committee was blissfully unaware of the EPA study that Mr. Bennett recently unearthed. The EPA says it marked the report 'do not cite or distribute' because it had not been given peer review. An EPA spokesman said Congress could have had it but no one asked.

"In other words, Congress didn't know because Congress didn't ask, and Congress didn't ask because Congress didn't know.

"This is silly. It is also dishonest.

"Had Congress known of the report's calculations, it would have acted sooner than it did to change the instruction manual."

The bottom line in the EPA report is a specific figure of 1,190 cancer cases caused by improper removal of asbestos. But when the figures are placed in perspective, the figure should be 3.71 times higher, or 4,414 cases.

About 35 percent of the 117,000 school buildings in the country or 32,860, contain friable, or loose asbestos. Approximately 20,000 had taken some action by the summer of 1984 to remove, encapsulate or enclose the material. Another 3,950 had adopted special operations and maintenance purposes defined "for the purposes of this analysis. . .as the marginal labor cost of a custodian to clean and inspect-containing friable materials for one hour a week per school.

The study focused primarily on 8,840 schools that as of the summer of 1984, had done nothing to abate asbestos. The projection of 1,190 cancer cases is based on the calculations for the 8,840 schools. The projection is equally valid for the entire universe or 32,860 schools or 3.71 times as many, which had taken no asbestos abatement action.

Simple multiplication of the 8,840 figure by the ratio of 3.71 yields the 4,414 total. It should be pointed out—as Ruckelshaus did in another context—that all these figures are highly speculative and theoretical. The base for all the calculations in the 105-page report was Selikoff's study of amosite workers in Patterson, N.J.

But assuming Selikoff's original figures and the subsequent extrapolations in the study are correct, the following data emerge:

A removal program in the 8,840 schools could prevent the cancer cases of 1,645 former students, 35 teachers, and 77 mainte- nance workers. But the prevention of those cases would have to be purchased at the price of other cancer attacks—caused by sloppy work procedures—on 1,045 former students, 35 teachers, 77 asbestos removal workers or 1,276 in all.

That's a difference—or net gain—of 550 cancer cases avoided in favor of removal.

But maintenance, by contrast, would spare 559 students, 19 teachers, and 26 asbestos maintenance workers, or 604 in all—104 more than removal. And not one life of one abatement worker would be cut short, or, for that matter, the lives of teachers, students and workers endangered by improper removal project.

Maintenance is the only one of the seven scenarios projected by the EPA cost/benefit analysis that does not demand the lives of asbestos abatement workers. The figures for the other six scenarios project as few as two or three to as many as 156 asbestos workers sacrificed in the theater of asbestos hysteria.

That's an argument for cost-benefit analysis. But there's a better one that doesn't just revolve around money, but the inherent stupidity of the absolutist or prohibitionist point of mind. Specifically, maintenance of asbestos would cost billions less than removal.

But that's not the real issue. A situation has been created that is even more dangerous than the human roulette that needlessly condemned the crew of the Challenger to death. A zero sum—Catch 22—scenario has been created in which everyone loses except the bureaucrats, the politicians and the media—and the lawyers.

Decisions are being taken to sacrifice some lives in favor of others. Can a government agency express a preference for a policy that prevents some deaths, even a substantial majority, at the cost of cost of others, even a few? Can prudent public health policies be constructed on the model of a roulette wheel?

A moral equation has to be squared. A moral balance has to be balanced.

But a moral fanaticism blinds those who compose *danses macabres* on computer consoles, debating like medieval theologians, not how many angels can dance on the head of a pin but how many carcinogens can cavort on a cursor.

They are members of what can be called an "environmental establishment," many of whom are deliberately exploiting fear of cancer to promote reform of American capitalism.

But the real results are wasted billions, needless psychological terrorization of children—and a scientific Inquisition sponsored and paid for by the federal government.

"Cancer is the cornerstone of the American industrial process," says Anthony Mazzochi of an organization called Parents Against Asbestos Hazards in Schools. He said that first when he was running for president of the Oil Chemical and Atomic Workers Union (OCAW).

Mazzochi, who was often reverently quoted by Paul Brodeur in articles for *The New Yorker* proclaiming the dangers of asbestos, wanted to use the OCAW presidency to form an independent labor political movement. This new labor movement, according to *In These Times,* a Socialist newspaper, "would emerge as an alternative to [the] developing bipartisan 'Party of Cancer.'" Mazzochi, who was strongly identified with the workers' safety and health movement in the '60s and '70s saw the '80s as a period of genetic confrontation.

Mazzochi is still promoting "genetic confrontation," but not with OCAW. He was defeated twice in bids for the union presidency. His rejection was partly due to his "corporate cancer" themes, for example, calling the chemical industry" mad bombers."

In recent years, Mazzochi has been found running around the country saying schools should be "closed down and evacuated" until all the asbestos is removed. And he has declared the country faces an epidemic of asbestos-related cancer "exceeding any. . .mortality. . .since the 1930s."

Fear of cancer, in other words, was—and is—being used as a political instrument.

This extreme view has been convenient for social reformers, according to Dr. John Higginson, former director of the International Agency Against Cancer, who originated the tag line that 60 to 90 percent of cancer is caused by environmental factors in the 1950s—and has been trying to explain what he meant ever since.

"Environment is what surrounds people—and impinges upon them," he told *Science* magazine, Sept. 28, 1979, "the air you breathe, the culture you live in. . .the chemicals with which you come in contact. A lot of confusion has arisen. . .because most people. . .have used the word 'environment' purely to mean chemical[s]"—including asbestos.

"The ecological movement, I suspect, found the extreme view convenient because of the fear of cancer. If they could possibly make people believe that cancer as going to result from pollution, that would facilitate the cleaning up of the water, that air, whatever it is.

"You asked me," he told his interviewer, "were people dishonest? I don't think some people were dishonest.

"People would love to be able to prove that cancer is due to the general environment or pollution. It would be so easy to say 'let's regulate everything so that we have no more cancer.' The concept is so beautiful that it will overwhelm a mass of facts to the contrary."

Concepts that defy reality can, in all too many instances, define political issues when scientists become political. And that has become more and more the case.

"While much is known about the science of cancer, its prevention depends largely, if not exclusively, on politics," Samuel S. Epstein, author of *The Politics of Cancer,* concluded, "This then is the message of the book."

Politics, however, and science are all too often incompatible. That was true when the Inquisition condemned Galileo to silence for proclaiming the sun, rather than the earth, was the center of the universe. It was equally true when Josef Stalin sanctified the "creative biology" of Trofim Lysenko in the hope it would create a new Socialist agriculture and superior Soviet human beings.

Both were not only tyrannies, but also failures, because both tried to substitute dogma for reality. And fanaticism is always the mortal enemy of science, as Claus and Bolander observed:

"[I]f Lysenkoist biology had merely represented concepts about the workings of nature which were different from our own we should have critically evaluated its results, and integrated them into our own science or rejected them as unacceptable purely on their merits.

"Even if the results had no practical value, we might have cherished 'creative biology' for its invigorating novelty. However, because the Lysenkoists not only presented new concepts but also took upon themselves a charismatic mission—and blessed by the [Soviet] system whose ideology they served—felt no trepidation at stamping out their adversaries through innuendo, intimidation, denunciation, or falsifications, they lost their claim to the title of scientists, pursuers of truth and furtherers of human understanding.

"By the same token, if the environmentalists were to content themselves with the presentation of new ideas which, although unacceptable within the terms of our present concepts about the workings of nature, might have in them some kernels of truth, we should cherish their insight and courage in bringing these problems to the attention of an all-too-conservative scientific Establishment.

"But when, in their ideological fervor, they employ tactics of innuendo, intimidation, and falsification, they are no better than their Soviet colleagues, for they prostitute science by subjecting it to a zealot's cause."

There are lies, damned lies, and statistics.
Anonymous

10 Numbers Games

On September 11, 1978, Joseph Califano, then Secretary of Health, Education and Welfare, proved once again the truth of Mark Twain's observation that a lie can get halfway around the world before the truth gets its boots on.

Califano shocked an AFL-CIO conference on occupational safety and health by announcing more than 2,000,000—two million—premature deaths would result from asbestos during the next three decades.

He further announced:

● "Five million American men and women. . .breathe significant amounts of asbestos fibers each day.

● "Seventeen percent of all cancer deaths in the United States each year will be associated with previous exposure to asbestos."

That would make the grand total of asbestos related deaths 6,300,300 in thirty years time, three times Califano's own estimate. Even the 2,000,000 figure was almost twice as high as the 1,200,000 projected earlier.

Dr. Irving Selikoff, up until then the most widely quoted asbestos authority, had estimated 40,000 excess deaths a year, 120,000 in 30 years, from asbestos a few weeks earlier than the Califano speech in testimony before an Occupational Safety & Health Administration (OSHA) public hearing considering a "generic" cancer control proposal. Five years later, Selikoff sharply scaled his estimate down to 8,200 cases-a-year, which would be only 246,000 cases over the next three decades. Yet Dr. Malcolm Ross of the U. S. Geological Survey has established no more than 520 actual cases were reported during what are generally accepted as the peak years for asbestos deaths, the 1970s, for a total of 15,000 cases.

The Califano numbers came out of a mimeographed paper titled "Estimates of the Fraction of Incidence in the United States," which estimated 20 to 38 percent of all cancer was attributable to occupational exposure. The National Institute of Occupational Safety and Health (NIOSH), OSHA's research arm, had earlier disputed those figures at an Occupational Safety and Health Administration (OSHA) public hearing on the proposed generic standard for regulating all carcinogens. NIOSH instead went along with the generally accepted estimate within the scientific community of three to five percent.

The "OSHA Estimates Document," as it came to be called, was a remarkable paper in many ways. There were two versions. The first, a "draft-summary" with a release date for the media, was attributed to the National Cancer Institute (NCI) and the National Institute of Environmental Health Sciences (NIEHS). The second contained a list of nine "contributors" listed in "'alphabetical order'—a point explicitly made on the first page the paper itself," as Edith Efron observed in her epochal book, *The Apocalyptics: Cancer and the Big Lie.* It had metamorphosed into an imposing study adorned with the names of some of the most important scientists in America's health agencies.

"The study was strange in one other respect," Efron observed. "It never had been published in a scientific journal. It had simply rolled off a government mimeograph machine right into the labor movement after being 'publicized'—the word 'publicized' used unselfconsciously by the Chief of OSHA's Media News Service in a letter written to inquiring citizens."

The document was curious in another respect. It was introduced directly into the record of the public hearing on the "generic" cancer policy by Dr. Eula Bingham, then assistant secretary of Labor for OSHA, a few days after Califano's speech. By doing so, the action precluded any effort by opponents of the policy to cross-examine the "contributors."

And the paper had "no responsible author," as Efron observed, "since by scientific convention, the alphabetized list of names meant that no individual among the seven was taking actual responsibility for the data in the paper."

The data was vehemently disputed within the scientific community, but is still often quoted by unions, environmental groups, and the media. But questionable as the figures are—and they certainly illustrate the fact that the computer has given new credence to the old maxim about the types of lies: lies, damn lies, and statistics—their evolution, more importantly, demonstrates how environmental politics has hopelessly compromised the fairness and objectivity of the rule-making process at agencies such at OSHA and EPA.

The significant question raised by Efron still needs to be addressed: Where had these pseudo-authors come from?

The answer to that question demonstrates how environmental advocates have turned the regulatory process into a moral equivalent of war against industry. It is a war fought with statistics rather than bullets, a numbers game. But, as in all wars, the first casualty was been truth in what has been called the civil war between "the lawyers vs. the scientists."

And the first victory was the creation of an "Environmental Establishment," composed of lawyers, scientists, politicians and their flunkies in the media, all determined to make the world safe for *New Yorker* readers.

On June 28, 1979, an article titled "OSHA: Cancer and the Politics of Fear," was printed in *The Congressional Record,* pages H17256-17260. The introduction was prefaced by the comment: "The OSHA proposal leads into one of the most peculiar stories of our times—peculiar on at least two different levels. First, there is a serious scientific question concerning the validity of standards OSHA is promoting. Second, there is the issue of how such dubious standards could be adopted by the federal government despite the paucity of evidence on their behalf. One is a matter of scientific fact, the other a matter of politico-legal maneuvering. Both deserve the most exacting scrutiny by the media and the Congress."

Both have received such scrutiny in the scientific literature, in books published by eminent political analysts as well as social interpreters, and among some key members of Congress. But with the exception of one series of articles that I wrote for the *Detroit News,* media attention has been only sporadic, without the systematic attention needed to prevent further waste of human life and resources.

That will change only after the story of what has happened has been pieced together, a story of political ambition, intellectual pride and social arrogance—if not legal corruption.

The story begins with three men at EPA shortly before and after William Ruckelshaus outlawed the pesticide DDT for "political, albeit with a small 'p' reasons." Two of the three were lawyers, Anson M. Keller and Anthony C. Kolojeski. The third was a scientist, Dr. Samuel Epstein, who had played a key role in the 17-month proceedings which had led to the refusal of the administrative law judge in the case, William Sweeney, to ban DDT.

Sweeney's conclusions were:

"DDT is not a carcinogenic hazard to man. DDT is not a mutagenic (genetic) or teratogenic (fetal) hazard to man.

"The uses of DDT under the registrations involved here do not have a deleterious effect on wildlife."

Sweeney conceded, "the evidence presented demonstrates a continuing need to pursue the truth as a carcinogen for humans." But, he added: "Really, the fact that DDT had NOT [emphasis his] been proven NOT to be carcinogenic to man is a logical basis for advocating a complete ban on all future uses of DDT"—a repudiation of a classic example of the impossibility of proving a negative argument employed by the environmentalists. Sweeney said he "gave a lot of weight" in the decision to the testimony of Dr. Jesse Steinfeld, the surgeon general at the time and a widely respected cancer researcher.

Sweeney gave considerably less weight to the testimony of Dr. Samuel Epstein, who, as the principal witness for Environmental Defense Fund, had argued DDT is carcinogenic at any level of exeposure. The authors of the most extensive history of the DDT hearings, *Ecological Sanity,* George Claus and Karen Bolander were directly critical of Epstein's experimental methods:

An "experimental technique is worthless unless it is properly executed," they wrote. "Epstein himself is a talented and well-trained research scientist. However, in the last few years he has taken an extremist stand on environmental chemicals and become something of a crusader. While the quality of this particular paper (tests of suspected carcinogen in mice) may not be typical of all his work, its sloppiness, imprecision, and inconclusiveness are indicators that when scientific researchers take positions against environmental threats, they may become so anxious to demonstrate that one or another chemical is dangerous that the quality of their experimental work deteriorates."

Epstein, who wrote a book, *The Politics of Cancer,* published in 1978, was certainly anxious to prove many chemicals are dangerous. The book was described in a *Washington Post* review as "compendium of all the information on the 'environmental' causes of cancer that is probably going to serve as a bible for the forthcoming crusade to drive all carcinogens from the workplace and the world."

Epstein himself transcended his presumably scientific objectivity to announce, as the conclusion of *The Politics of Cancer:* "While much is known about the science of cancer its prevention depends largely, if not exclusively, on political action. This then is the message of the book."

But politics does not wear the white coat of the scientist—and that was to become abundantly clear in the following years as what was becoming an Environmental Establishment gave signs of becoming an environmental racket.

Epstein's next great, and perhaps most significant, opportunity to cleanse the world of carcinogens came in assisting Keller and Kolojeski in their 1973-75 efforts to ban aldrin-dieldrin, a chlorine-based pesticide similar to DDT.

"The [EPA] litigation team," Epstein recalled in *The Politics of Cancer,* "was composed of Anson Keller and Anthony Kolojeski of the Office of General Counsel. This team started work in virtual isolation, as the Office of Pesticide Programs where the supposed scientific expertise on pesticides supposedly was still highly sympathetic to agrichemical interests quite apart from resenting the Office of General Counsel's access to the administrator [Ruckelshaus]. . . . The Office of General-Counsel was staffed by young, environment-minded lawyers. The bitter schism which developed, known as 'the scientists vs. the lawyers,' largely reflected the fundamental political ambivalence between environmental activism and traditional pro—industry conservatism."

The "failure" of EPA scientists, in Epstein's words to provide the evidence wanted by the lawyers, however, also "opened the door to go outside the agency." But the alternative proved "not to be so easy. Most university agricultural economists and entomologists receive research support from industry and were unwilling to take up the government position."

A majority of experts on toxicology and carcinogenics who were approached were either in a similar position or unwilling to take the time to help. Thus the government team and a handful of independent outside experts were organized by Epstein. That "handful of independent outside experts" were agreed on "nine cancer principles." The most important was: "The concept of a 'threshold' exposure level for a carcinogenic agent has no practical significance because there is no valid method of establishing such a level."

That, of course, is the guiding principle behind the Delaney Clause of the 1958 Food and Drug Act calling for a ban on all carcinogens in foods—regardless of the amount. And Kolojeski and Keller were successful in securing a ban on aldrin-dieldrin by invoking it. The clause to the 1958 Act may have made some sense then when it wasn't realized that substances such as arsenic, a carcinogen as well as the "deadly poison of the Borgias" Rachel Carson compared DDT to in *Silent Spring*, are essential elements needed to maintain human health.

But the concept behind Delaney has now been repudiated by the National Academy of Science. The Academy, reacting to Congressional concern over efforts to ban saccharin, has recommended the clause be amended to reflect the reality of dose-response relationships by classifying carcinogens as "high," "medium," and "low."

But Kolojeski and Keller argued then—and the environmental movement does today—there can be no safe threshold of exposure, "one fiber can kill" in the case of asbestos. "Foremost of the regulatory amendments in its offer of informed guidance is the Delaney Clause," Keller and Kolojeski wrote in their brief to the administrator. "In effect since 1958, this statutory prohibition constitutes the judgment of Congress that the addition of any amount of carcinogenic material to the food supplies, regardless of its purpose or value, is hazardous to public safety, and that a chemical which is carcinogenic in laboratory species constitutes an unacceptable risk of causing cancer in man, i.e. additives to the food supply which are carcinogenic in laboratory animals, must, perforce, be assumed carcinogenic in man."

Among the "independent" experts assembled by EPA were Dr. Umberto Saffiotti of the National Cancer Institute (NCI); Dr. Arthur Upton, then at the University of New York at Stony Brook; and Dr. Marvin Scheiderman of NCI. Dr. David Rall, the director of the National Institute for Environmental Health Sciences (NIEHS), did not directly advise on the pesticide cases, but in Epstein's words "has been involved in critical issues such as defending the Delaney Amendment from industry attacks, and in backing the proposed ban on saccharin."

Upton, who was one of the founding members of the Environmental Defense Fund along with Epstein, was to become director of NCI during the Jimmy Carter years. Rall, Saffiotti, Scheiderman and Upton were all to a play a significant—and alphabetically precise—role in the development of the "Estimates Document."

Keller, Kolojeski, Epstein and "the handful of independent scientists" had won the case for a ban on aldrin-dieldrin as well as a ban on another pesticide with similar chemical properties, heptachlor-chlorodane. But the victory was short-lived, for shortly thereafter, Congress forced EPA to reorganize. In Epstein's words, EPA's internal policies were revised "to exclude the possibility of initiation of further litigation by the Office of General Counsel and to place its responsibility, instead, largely in the hands of its Pesticide Regulation

Division."

The lawyers were no longer in charge, in other words.

It was time to move on. Keller left to become a special assistant to the assistant secretary of Labor for OSHA. Kolojeski formed a private consulting firm, Clement Associates. Keller, in his new role, was assigned, in Epstein's words, "top priority to promulgating standards for occupational carcinogens. . . . Determination of carcinogenicity in these proposals is generic, as in the Delaney Amendment."

Now the focus shifted from DDT, aldrin-dieldrin and heptachlor-chlorodane and other pesticides to asbestos which was to become the devil to be exorcised by the environmental movement.

Original studies done by Sir Richard Doll of Oxford in the '50s had established asbestos workers with 20 years exposure had ten times the lung cancer deaths of non-asbestos workers of the same age. Followup studies by Dr. Irving Selikoff of Mt. Sinai Hospital in New York found similar rates in the United States, but also noted cigarette smoking multiplied the risk 80 times over. Doll's subsequent findings on smoking made it obvious cigarette smoking was the principal cause of lung cancer—and Doll was to make that eminently clear during the subsequent clash over the estimates documents.

In the meantime, however, Epstein, disregarding the facts, had declared: "The massive human toll taken by asbestos is probably the single most important incentive to the development of coherent national policies recognizing preventive medicine as a major future component in the delivery of human health care."

But the facts were:

1. There was no massive human health toll taken by asbestos.

2. The coherent national health policies called for by Epstein were grounded in politics—not science—and should then and now be as critically and skeptically assessed by Congress, the media and the public as any other panacea for ills of society.

That meant the relationship between Kolojeski and Keller deserves attention. A *Wall Street Journal* article printed on October 11, 1978 did so, but, unfortunately, it was an exception that proved the rule that the media is apparently incapable of closely examining environmental regulatory agencies.

The *Journal* article read, in part:

"Many observers in Congress and industry believe OSHA's recent cancer hearings reveal a certain scientific close-mindedness at the agency. . . . [T]hey insist the hearings were dominated by executives of an outside consulting firm, called Clement Associates, Inc. That firm had been retained by OSHA to provide scientific expertise in evaluating testimony, finding expert witnesses and determining which substances should be regulated under the proposed cancer policy. . . . Clement Associates performed much of the work that would be been performed by officials of the National Institute of Occupational Safety and Health (NIOSH).

"Some NIOSH officials feel that OSHA, departing sharply from past practice, purposely excluded the institute from the process of shaping the cancer proposal. NIOSH officials weren't invited to testify at last summer's hearings, and they couldn't even get copies of the proposal until just before it was released publicly. 'NIOSH was completely bypassed,' says the institute's executive director, Edward Baier, 'To what purpose and for what end, I cannot say.'" The *Journal* quoted critics as saying "NIOSH was bypassed because OSHA officials didn't want to deal with an agency not always in sympathy with OSHA's regulatory outlook. . . .

"According to this view, OSHA sought the assistance of a more sympathetic organization, Clement Associates. OSHA's critics bolster their case by pointing out that the agency's top lawyer in the cancer-policy area, Anson Keller, once worked closely at the Environmental Protection Agency with Clement's president, Anthony C. Kolojeski. At the EPA, the two pressed for many of the same policies that are now incorporated into OSHA's anti-cancer proposal."

The *Wall Street Journal* did not go deeper into the relationship between Keller and Kolojeski. Had it done so, some interesting questions of intellectual and contractual propriety might have been raised.

In October of 1976, *The Congressional Record* of June, 1979, detailed what was called the "OSHA-Clement Sweetheart Deal." The account began with the description of the letting of a non-competitive $10,000 purchase order to Clement Associates from OSHA. The contract was to develop "Alternative Carcinogen Regulations." The contract was approved by Dr. Morton Corn, then assistant secretary of Labor for OSHA, now head of the industrial hygiene program at Johns Hopkins University.

"I knew nothing about the previous relationship between Keller and Kolojeski," Corn, who has since become one of the most notable critics of federal regulatory action on asbestos, said years later. "At the time, administrations were changing and I just wanted to make sure regulation of carcinogens proceeded more effectively than it had in the past. In fact, before leaving office, I wrote an extensive paper detailing what I thought should be done. Needless to say, it wasn't implemented as I hoped."

But quite a lot more was done, for a mere $10,000 purchase order by Clement Associates than Corn either wanted—or desired. For Clement Associates was prepared to do the following:

Scope of Work:

The contractor shall examine and furnish alternative approaches and policies for the regulation of carcinogens, and shall examine OSHA regulations and proposed regulations for carcinogens, providing recommendations for improvements in or alternatives to OSHA initiatives.

Statement of Work:

The contractor shall provide highly qualified professional staff with appropriate background to deal with the following tasks:

A. Survey of existing federal regulations for carcinogenic, mutagenic or tetragenic substances. Provide comparative analysis of regulations and policies and prove a crucial evaluation of effectiveness and problems of both implementation and enforcement of other federal regulations.

B. Develop prototype regulations for controlling employee exposure to carcinogens found in the occupational environment.

C. Provide position papers on defining carcinogenic agents and on OSHA regulatory policy for those substances found to pose a carcinogenic risk.

All of that—for a mere $10,000—seems like a solid investment in the health and safety of American workers. Unfortunately, we will never know what that $10,000 bought.

"No formal reports were asked of Mr. Kolojeski," David C. Zeigler, director of administrative programs for OSHA, wrote on November 3, 1978, in response to a Freedom of Information request. "As a matter of expediency, the input to the agency was in the form of briefings or discussions of issues, handwritten or typed opinions and summaries of data available on subjects assigned by OSHA and 'markup of' the draft documents prepared by OSHA.

"The documents were used in general as source documents for OSHA staff, and as appropriate some sections of these documents were incorporated in early drafts of the proposed cancer policy. These early drafts were continually reviewed and revised by OSHA staff and in January of 1977, a preliminary draft was publicly released."

Clement, having played a leading role in developing the preliminary draft, now won a competitive bid on June 22, 1977, to provide technical assistance to OSHA during the public hearing scheduled for the summer of 1978. This time the financial terms were considerably different. Under the terms of the original purchase order, the maximum charged was $15-an-hour. The new contract called for $43-an-hour for a project manager, $48.25 an hour for senior scientists and $14.50 for junior analysts. On the basis of a forty hour week, that runs to $85,000 a year for the manager and $98,000 for the scientists and $27,000 for the junior analysts. Those weren't bad wages in a time period when comparable federal salaries were one-third to one-half those respective amounts.

Obviously, at that rate, the total bid figure on the contract, $65,300 would be quickly eaten up. That little fiscal problem was to be remedied very soon. On July 26, 1977, one month after the contract was awarded, OSHA's technical representative, Dr. James Vail, a career civil servant and a scientist, was replaced. His successor was Special Assistant Keller.

Three weeks after that, on August 18, 1977, the amount of the contract was tripled to $213,000. Between then and November of 1978, Keller signed off on invoices totaling almost twice that amount, $400,000. Mr. Zeigler, in a letter dated January 19, 1979, said: "We anticipate that further payments will be made for services rendered as provided for in the contract." Unconfirmed reports place the total paid as being well in excess of $1 million.

Clement Associates carried out its responsibilities to OSHA under the contract through two task orders. The first order called for screening 1,500 suspect carcinogens identified by NIOSH.

On April of 1977—about three months before Clement was awarded the OSHA contract—the firm secured another contract with the National Science Foundation to provide technical assistance to the Interagency Advisory Committee established by the Toxic Substances Control Act. The Act, under which asbestos is regulated and may be banned by EPA, contains an interesting omission—a definition of toxicity, yet again another implicit acceptance of the no-threshold theory.

That Clement Associates contract, too, was to screen the 1,500 suspect carcinogens identified by NIOSH. The first phase of the multiphase contract was for $142,793 for 3,140 hours of work—$45 dollars an hour—to be completed by June 23, 1977, one day after Clement Associates was awarded the OSHA contract.

In other words, Clement would presumably have already completed a survey paid for by another federal agency that could be simply replicated for OSHA. The Clement Associates proposal to OSHA mentioned the fact it had already been retained by an InterAgency Committee to do essentially the same work: "Although the criteria to be used in setting priorities for testing under the Toxic Substances Control Act are not identical for regulation under the Occupational Safety and Health Act, the overlap is sufficient to give Clement Associates valuable insight into and experience in this task," the company's proposal read.

The task was hardly an onerous one. The primary source for both projects was the list of suspect carcinogens updated and compiled annually by NIOSH then. The data is obtained by simply skimming the scientific literature, a task performed primarily by graduate students.

No attempt, whatsoever, was made to appraise or assess the quality of the research.

The list is published in book form and also fed into a computer. The information can be bought from NIOSH in either form. Clement Associates "like everyone else have requested access to our computerized information service which is called NIOSH-TIC," Dr. Geoff Taylor testified at the OSHA hearing on generic cancer policies.

The generic cancer hearing was suspended two months before the Estimates Document was released and never resumed. Consequently, representatives of affected interests were never able to ask whether NIOSH could have performed the task done by Clement for a fraction of the cost.

Nor was anyone able to cross-examine Rall, Saffiotti, Schneidermann and Upton, whose names were listed last on the list of "contributors" to the now notorious document under a curious arrangement noted by Efron. Yet, in order of importance within both the scientific and environmental establishments, the four would have normally ranked first. Rall was—and is—the head of NIEHS. Upton was director of the NCI.

Saffiotti, NCI's top career researcher, has been described by Efron as having "perhaps more than any other contemporary scientist translated the problems of carcinogenesis into an organized body of theory," and "one of the founding fathers of the contemporary crusade against industrial cancer." Saffiotti's philosophy is perhaps best expressed in a 1976 statement: "[F]or prudent toxicological policy, a chemical should be considered guilty until proven innocent."

Scheiderman, an NCI statistician and fervent advocate of the no threshold theory, first announced the United States was in the throes of a cancer "epidemic" in testimony before the 1978 OSHA hearing. He spoke out two years before actual data, which might have supported his prediction was assembled by NCI. As events turned out, the data was never published in a scientific journal or peer-reviewed. And Schneiderman himself recanted his earlier prophesy in an news story published in *The New York Times*. That was after Doll and Richard Peto of Oxford published an article in the *Journal of the National Cancer Institute* and later as both a book, *The Causes of Cancer,* and a report for the Office of Technology Assessment, which provides Congress with independent scientific advice. Doll and Peto concluded the NCI data was "grossly discrepant with any generalized increase other than that due to tobacco."

The Causes of Cancer contained a comprehensive and withering attack on the estimates document "as so grossly in error that no arguments based even loosely upon them should be taken seriously." Doll and Peto continued, in Efron's words "with scorn and language. . .of a type never before directed at American public health agencies," as follows:

"It seems likely that whoever wrote the OSHA paper [it has a list of contributors, but no listed authors] did so for political rather than scientific purposes, and it will undoubtedly continue in the future as in the past to be used for political purposes by who wish to emphasize the importance of occupational factors. . .[in] many newspaper articles and much scientific journalism. However, although its conclusions continue to be widely cited the crucial parts of the argument for these conclusions have perhaps, advisedly, never been published in a scientific journal nor in any of the regular series of government publications. Unless they are, with proper attribution of responsible authorship, we would suggest that the OSHA paper not be regarded as a serious contribution to scientific thought and should not be cited or used as if it were. Furthermore, any suggestions which derive directly on indirectly from it that 20, 23, 38 or 40 percent of cancer deaths are, or will be due to occupational factors should be dismissed."

Doll and Peto's attack was far from the only one. Dr. Cuyler

Hammond, who had collaborated with Selikoff in the study of asbestos workers with the highest exposures, insulators, observed the document used "rather tricky arithmetic." Philip Abelson, editor of *Science* attacked the "opportunists" alleging the existence of a cancer epidemic. Professor Rene Truhaut of the University of Paris, speaking before the International Conference on Chemical Toxicology, called the study a "scandal," and supported the British Royal Society's estimate that only one percent of cancer was of occupational origin.

Dr. Irving Kessler, a leading epidemiologist, worried that the discipline of cancer study could degenerate into "mindless statistical manipulation. . . . I'm very much concerned about the political implications of enunciating cancer risks based on speculation, with pitifully little empirical evidence." The British science journal *Lancet* bemoaned "such a sorry product could appear under such distinguished names."

Dr. John Higginson of the International Agency Against Cancer (IARC), whose observation in the 1950s that environmental factors accounted for 80-90% of cancer was caused by environmental factors was systematically distorted by EPA's Office of Public Awareness for years, produced a new study published in *Preventive Medicine*. The Higginson article was peer-reviewed by some of the world's most distinguished cancer researchers, including Dr. Johannes Clemmesen of Denmark, a founding father of the environmental theory of cancer and its first historian.

Based on data from the industrial city of Birmingham, England, Higginson concluded: Tobacco might account for 30 percent of male cancers; diet, sexual behavior, 30%; sunlight, 10%; tobacco/alcohol 5%; radiation, 1%; and medical, 1%,; leaving 15% unaccounted.

Asbestos, as a cause of cancer, was not even mentioned

Saffiotti, one of Epstein's, Keller's and Kolojeski's band of "independent scientists" came under personal attack by his scientific establishment colleagues on September 26, 1978 at a meeting where he was the only one of the "contributors" to the "Estimates Document' who accepted an invitation to appear. Peto, who was present, observed the asbestos figures cited in the OSHA document were "possibly 1,000 times" higher than the Selikoff-Hammond data. "It's comical," he added, after earlier observing Upton and Rall "still seemed more ready to defend it [the document] than to repudiate it." Dr. John Cairns, then head of the Mill Hill Laboratory, observed several parts of the document were "manifestly silly," and anyone who could perform calculations could "see how stupid it is." Saffiotti took offense at Cairn's phrase "manifestly silly" and defended the document in the following muddled terms:

"And this I emphasize again is just an early stage of this analysis and because of curious circumstances became much more of a public document than it was originally planned to be. I hope this [meeting] at least will stimulate those whose names are listed in this document with facts, data. . . ."

But Saffiotti never addressed the "curious circumstances," as Efron observed, "which led a document without authors, 'facts,' and 'data,' to be funnelled through the mouth of HEW and directly to the press without peer review." And his colleagues failed to follow up with questions which probed into the "political phenomenon," in Efron's words, "which had aggressively misinformed the country and gratuitously terrorized the labor movement. Instead, only statesmanlike sorrow was expressed about the impact on the scientific community of "a statement which is going to fall apart. . .and damage the enterprise which so many of us are concerned with."

None of the concern expressed at that meeting, however, was communicated through the media to the public. And Rall just a few days after attending that meeting of his peers, went to an unprecedented "seminar" for the media in Chicago—all expenses paid for by OSHA and the federal government—where he defended the document.

The environmental establishment had co-opted the scientific establishment and conned the media.

Now everyone was playing the numbers game and no one was asking hard questions—yet.

But the questions—and some solid answers—were to come.

*One belief, more than any other, is responsible for the
slaughter of individuals on the altars of great historical
ideals. . . .This is the belief that somewhere in the past, or
in the future, in divine revelation or in the mind of an
individual thinker, in the pronouncements of history or
science, or in the simple heart of an uncorrupted man,
there is a final solution.*
                                                    —Sir Isaiah Berlin

11 Beyond Hysteria

It had to happen.

It was as inevitable as a Greek tragedy—or farce.

"Farce is the essential theater," said British actor Gordon
Craig (1872-1966). "Farce refined becomes high comedy; farce
brutalized becomes tragedy."

The Greeks always displayed the masks of comedy and tragedy
together, for they knew life repeats itself, first as tragedy, then as
comedy, and, ultimately, as farce.

Dwight Welch is a scientist and a professional career employee
at EPA. He is also the president of the EPA Local 2050 of the National
Federation of Federal Employees (NFFE), the agency's professional
union, composed primarily of scientists.

The union was formed largely because, in the words of a former
union president, J.W. Hirzy, writing in *Environmental Law Reporter*
of February, 1990, "Somewhere along the line a decision was made
to limit the number and caliber of scientists employed by the agency.
Rule writing was emphasized over scientific investigation. In effect the
science element of EPA became not only subordinate but *subservient*
[emphasis Hirzy's] to the legal element."

EPA had almost 16,000 employees as of May, 1989, according
to the U.S. Office of Personnel Management—15,615 to be precise,

almost as many as the Labor Department's 18,342 and the Energy Department's 17,121, considerably more than the Department of Housing and Urban Development's 13,454, not to mention the Education Department's mere 4,694.

Yet, according to Hirzy, only one scientist at EPA—a chemist—has a GS-15 rating, the top level in the federal career service with a pay scale of approximately $61,000 to $73,000 a year. There are no GS-15 epidemiologists. According to Hirzy they were "fired *en masse*" during the 1980s. Nor are there any GS-15 toxicologists, nor pharmacologists, nor biologists, nor climatologists, nor botanists.

Nor is there any career path for research scientists at the agency. By contrast, the U.S. Geological Survey with approximately 10,000 employees has, according to Dr. Malcolm Ross, "hundreds of GS-15 scientists." And promotions are based largely on professional achievement based on publication in outside peer-reviewed scientific publications.

By direct contrast, the management structure established in EPA right from the beginning was composed of"bureaucratic fiefdoms" in Hirzy's words, run by lawyers. Dr. Samuel S. Epstein, in his 1978 book, *The Politics of Cancer*, refers to the "bitter schism" commonly known within the agency, "the lawyers vs. the scientists."

This "law over science" operating philolosopy reflects the fundamental difference between the ethics of lawyers and the ethics of scientists. A. James Barnes, former general counsel for EPA, said, "Lawyers generally help policymakers go where they want to go." Hirzy observed, "the scientist's duty, on the other hand, is to uncover nature's secrets and publish his or her findings, irrespective of any 'client's' desires."

The conflict is as old as the Greeks. Philosophers—scientists—seek truth for itself, as Socrates observed. Sophists—lawyers—"make bad the better cause."

Welch was aware of that bitter irony when he wrote an open letter to EPA Administrator William K. Reilly in November of 1990 following an independent inspection by the firm of Booz-Allen of an asbestos removal project conducted by EPA in an EPA-leased building in Crystal City, Virginia. The building accommodates numerous EPA scientists, many of them reporting, as Welch does, to a manager whose only formal academic qualifications are a bachelor's degree's in English. Compounding the irony, "union-led" protests, in Hirzy's words, had played a key role in formulating EPA's original asbestos-in-schools rule.

Welch wrote:

This is an "asbestos removal project, which is, on its face, sloppy, shoddy, apparently illegal, and may be endangering the health, and ultimately the lives, of the occupants."

Strong words. More followed:

● Assurances that work specifications would be made available to EPA employees and union officials were "a pack of lies."

● Officials responsible for the removal project were described as "scientifically illiterate and without conscience."

● EPA has subsequently "lied repeatedly to its employees over this asbestos removal work."

● A provision in the work specifications calling for "no visible" asbestos is "by itself. . .garbage and arbitrary."

● EPA "may be casting its employees into the throes of uncontrollable revulsion."

EPA is, in other words, a hypocrite, a liar, and a political bungler as expressed in a blunt warning written in a September, 1990, letter from Loree Murray, president of the EPA local of the American Federation of Government Employees (AFGE), representing the agency's pink and blue-collar workers:

"Remember, EPA is a model for the nation. Improper management of this project could be very embarrassing at the very least."

Welch called for the "immediate firing" of EPA and General Services Administration (GSA) officials for "their criminal disregard for human health and well-being."

The project, in fact, gave every indication of being a a classic example of what EPA Administrator William K. Reilly had warned against in his June, 1990 speech, "Asbestos, Sound Science and Public Policy, Why We Need a New Approach to Risk:"

"[A]n asbestos situation that may have posed virtually no risk at all could, if this removal is not properly conducted, turn into a fairly significant health risk.

"Furthermore, schools are not the only buildings where unnecessary asbestos removal is being carried out; we've also learned that a number of commercial buildings have also undertaken expensive and, from a health standpoint, probably unnecessary removals.

"In some cases, it appears that mortgage bankers are *requiring* [emphasis Reilly's] that asbestos be removed before approving loans secured by the property, because they're worried about property devaluation. In other cases, building owners are concerned about the expense of liability insurance or the cost of legal claims if an occupant of the building develops an asbestos-related disease."

The scenario sketched by Reilly seems to fit this situation precisely. The privately-owned building is approximately 20-years-old, and, according to a former worker on the removal project, was being re-financed.

Despite a cost of $5 million, about as much as the original cost of construction, the work was undertaken partially because the primary source of asbestos was amosite in ceiling tiles. Amosite is accepted universally within the scientific community as posing significantly greater adverse health effects than chrysotile.

Ironically, only five percent of the 730,000 public and commercial buildings suspected of containing asbestos, 3,650, are believed to hold amosite or crocidolite, the two major amphibole varieties which account for most asbestos-related disease. The other 95% contain chrysotile.

But irony compounded irony, for the results of the Booz-Allen survey provided clear evidence that the removal project disturbed unknown sources of chrysotile. Of 24 samples of dust taken, only one contained measurable amosite, 1%. But 14 of the 24 contained chrysotile at measurable limits, eight at 1%, four at 2%, one at 3% and one at 4%.

The sample at four percent was found inside a rotating file cabinet, the three percent sample on the top of a window, the two percent samples were on floors or the top of files, the one percent samples on floors, fireboxes—and in or near air ducts and louvers. All could easily be "entrained," released back into the air-conditioning and heating system in the totally enclosed building.

As a further irony, according to Welch, EPA would not agree to conduct air sampling. The dust was simply collected in vacuum cleaner bags, instead of being blown out into the ambient atmosphere.

"They said that would be too dangerous, even with moon-suits," Welch said.

"According to some federal regulations, the one percent levels do not constitute a problem," according to Malcolm Ross of the U.S. Geological Survey, "But the two, three and four percent measurements indicate it was a godawful job that never should have been done in the first place."

That, of course, was the central message in the so-called "Green Book" almost surreptitiously released by EPA in September of 1990:

"Removal is often *not* a building owner's best course of action to reduce asbestos exposure. In fact, an improper removal can create a dangerous situation where none existed before.

"By their nature, asbestos removals tend to elevate the airborne levels of asbestos fibers. Unless all safeguards are properly applied, a removal operation can actually increase rather than decrease the risk of asbestos-related disease."

The "Green Book" was released the week before the letters from Welch and Murray were sent.

The book reiterated in official form the message Reilly was apparently trying communicate in June:

● "Although asbestos is hazardous, the risk of asbestos-related disease depends on exposure to airborne fibers. ... [A]t very low levels, the risk may be negligible or zero,

● "Based on available data, the average airborne asbestos levels in buildings seem to be very low. Accordingly, the health risk to most building occupants also appears to be very low. . . . [M]ost building occupants, (i.e. those unlikely to disturb asbestos-containing materials) appear to face only a very slight risk, if any, of developing asbestos-related disease."

That was the message EPA was trying to send, Reilly said.

"EPA has been trying, especially within the last few years, to emphasize the importance of managing asbestos 'in place,' wherever possible. We've stressed the approach because the unnecessary removal of asbestos may actually pose a *greater* [emphasis Reilly's] risk than simply leaving them alone, as long as the materials are undisturbed and unlikely to be disturbed.

"As is true with any hazardous substance the mere *presence* of asbestos poses no risk to human health; only when asbestos fibers are released into the air and breathed into the lungs do they become a health risk."

The effectiveness of the communications efforts can perhaps best be best judged by the comments of EPA's own union leaders about the "Green Book" in December of 1990:

"I'm not sure what that is," Welch of the NFFE said.

"I'm not sure I'm aware of that," said Kirby Biggs of the AFGE, "I may have seen a memo on it. But I don't think I've really heard about it until now."

In his June speech Reilly announced he had "commissioned" his assistant administrator for communications, Lew Crampton, to conduct a "major management review of our asbestos communication effort."

Six months later, Welch said: "Well, he [Crampton] must be doing a real great job. But I've never heard from him, never mind talk to him."

A final draft of Crampton's report, circulating within the agency in the summer of 1991, admitted–as much as bureaucrats can admit error–that the agency had "inadvertently contributed to the [public] confusion by issuing evolving messages over time." Early guidance books had stressed "removal is the only 'permanent' solution."

The so-called "Purple Book" issued in 1985 "offered a new twist to an old message–improper removals may be an even greater hazard than if undamaged asbestos were left alone." The Green Book, issued very quietly if not surreptitiously in 1990, "strongly emphasizes the hazards associated with improper removals and stresses that in-place management may often be a school's best asbestos alternative."

The "communications review" document admitted, "Some parties outside EPA have characterized the Green Book as a 180-degree shift in agency policy." But the "communications" document also contained what can only be described as a flatout lie, a claim that the Mossman *et al.* articles in *Science* and *The New England Journal of Medicine* "marked the first new scientific thinking on asbestos hazards." The lie is particularly damning because Crampton had been given a copy of an almost complete draft of this book in the summer of 1990.

EPA is, at worst, a "Ministry of Truth," in the phrase coined by George Orwell for his novel *1984*–an "Office of Public Awareness," peddling panic and fear instead of imparting truth and knowledge.

At best, "this place is just a welfare program for the upper middle class," according to Dr. Rufus Morison, chief steward of the NFFE. "How can they tell people what to do about asbestos when they don't know how to handle it themselves?"

"This is heartless ineptness," said Biggs of the AFGE. "The way EPA treats its own employees, I can understand why the public gives us grief all the time."

The fundamental problem is that a government agency, protected by the ancient legal doctrine that "the King can do no wrong," cannot be sued without its consent. Therefore, the United States government cannot be held legally accountable for the direct and indirect actions it took in failing to adequately protect the safety and lives of those directly affected by asbestos-related disease, World War II shipyard workers.

However, that abrogation of responsibility has been addressed in a decision of the U.S. Court of Appeals for the Federal Circuit in Washington, D.C.

By a 2-1 decision handed down in July, 1990, the court ruled, as reported by Associated Press, "three asbestos makers can sue for compensation paid to sick workers at U.S. Navy shipyards. . . .

"The government to date has not yet admitted responsibility or agreed to pay compensation for illnesses contracted by workers at Navy shipyards where asbestos was used, predominantly during and after World War II."

The decision is expected to prompt dozens of companies to file suits to force the government to share in payments already exceeding $1 billion for asbestos-related disease affecting tens of thousands of workers.

"So far only our clients have paid the tab," said Joe G. Hollingsworth, an attorney representing two of the three companies involved in the suit, UNR Industries, Inc. and Eagle Pitcher. "The government has escaped liability by virtue of technical arguments."

Federal courts are also trying to consolidate an estimated 600,000 suits clogging U.S. court dockets. The asbestos processors have argued that the government should be held liable because it allegedly failed to tell them about asbestos' health hazards and because it was the government's responsibility to maintain a safe workplace.

"The government specified products with asbestos in them, the government controlled working conditions and the government in the view of many people has the primary responsibility for what happened," said, a Newark, N.J. counsel for several asbestos processors.

If the government itself cannot or will not bring itself to admit it can be wrong, individual government officials may.

On July 12, 1989, the very day a final ban on asbestos was announced by EPA Administrator Reilly, to be effectively abjured less than 11 months later, a remarkable event occurred.

A former federal official publicly admitted the government was wrong.

Grover Wrenn had been the director of health standards at the Occupational Safety and Health Administration (OSHA) when the original standards regulating exposure to the mineral had been promulgated in the 1970s. And he had been the direct supervisor of Anson Keller, the OSHA special assistant for regulatory affairs, when the agency attempted to set a generic standard for all carcinogens.

That effort, which detonated a worldwide scientific controversy, had been orchestrated through a private company, Clement Associates. The president of the company, Anthony C. Kolojeski, had been an associate of Anson Keller at EPA in efforts to successfully ban a number of pesticides. Wrenn went to work for Clement as a vice president after OSHA's generic carcinogen policy collapsed.

On July 2, Wrenn made a public statement under the auspices of the Asbestos Information Association of North America.

This is what he had to say:

> When I first became acquainted with asbestos at OSHA in the mid-1970s, there was a serious need to enact strict regulatory controls to protect workers.... [H]istorical asbestos exposures in the workplace were very high, sometimes up to 100 or more fibers per cubic centimeter (fibers/cc).
>
> The initial OSHA permissible exposure limit, issued in 1971, required exposure levels to be below 12 fibers/cc. That standard has progressively moved downward to the current PEL (permissible exposure limit) of 0.2 fibers/cc–1% of typical historical levels.
>
> The (current) OSHA standard is the tightest in the world, indeed five times higher than the PEL recommended just this April by a group of international experts under the auspices of the World Health Organization (WHO). In addition, OSHA has mandated numerous work practices to further protect workers in this country who work with asbestos.

I visited just last month one of the two remaining asbestos-cement (A/C) pipe plants in this country. Exposures to workers in that plant are well below the OSHA standard. I only wish EPA's policymakers had been with me to see how effectively asbestos exposures are controlled in current A/C pipe manufacture.

As an industrial hygienist by training, I am a member of a profession whose purpose in life is to protect workers from risk by the recognition, evaluation and control of workplace hazards. But, like all industrial hygienists, I am aware that the elimination of all risk is impossible.

The public may not realize it, but it is well documented that the lifetime risk of death from work-related accidents and diseases ranges, for various occupations, from 1 to 20 per 1,000 persons. Thus, when a health regulation reduces risks to the lower end of that range, the regulation is an important achievement in assuring that significant risks do not exist in that occupation.

The risks EPA estimates for asbestos workers, given the OSHA requirement that exposures be below 0.2 fibers/cc are in the lower end of that range. Those estimates are based on theoretical models of the risk of asbestos; in fact, many scientists believe there is no observable risk at such exposure levels.

But, even if EPA's estimates are believed, the risks it is regulating here by banning use of asbestos are not unusual nonoccupational risks; certainly these risks do not warrant the extraordinary regulatory response represented by this ban.

Society needs to control toxic substances effectively to assure they do not pose significant risks. But asbestos is already being regulated stringently. Eliminating all toxic risks is an impossibility, and EPA's attempt to single out asbestos to achieve that goal makes little sense when substitutes are much more expensive.

Asbestos is a valuable mineral that performs an important role in products that are less expensive and/or more effective. EPA's own assessment of the additional hundreds of millions of dollars that it will cost society to employ the substitutes confirms that. The regulatory goal should be effective control of exposures—not bans on use.

The importance of considering the costs of asbestos substitutes is especially strong if those substitutes themselves pose risks to health and safety.

Regardless of the auspices, those words just make sense, scientific as well as common. The basic position will be repeated in arguments before federal appeals courts, probably all the way to the Supreme Court, seeking an overturn of EPA's asbestos ban.

If, as Charles Dickens observed, "the law is an ass," it follows that those who make laws can be and often are asses. A Supreme Court decision could overturn the asbestos ban. But it is unlikely the court will undo the damage done by the Asbestos Hazards Emergency Response Act (AHERA) or repeal the law itself.

That will take political action through an informed and justifiably angry citizenry. The information necessary to inform them is available through what was until October of 1986 a little known and previously uninovoked provision of the Toxic Substances Act. The provision was invoked in a legal challenge to the agency's announced intention to place a total ban on asbestos. The provision forced witnesses for the agency to submit to cross-examination under admonishment of possible legal consequences for failure to tell the truth.

The experience was unprecedented in American regulatory practice, although common in British Commonwealth countries where a system of Royal Commissions has been used over the years to assess and make recommendations on matters of safety and health. The British system has the advantage of relative objectivity, since the commissions are convened only to assess the merits of proposed regulations and go out of business immediately thereafter.

The three members of a Royal Commission are entirely independent with no professional stake in multiplying regulations. They are judges, not prosecutors, not mid-level bureaucrats assessing a problem with an eye on top-level jobs.

Thus, these commissioners have a certain degree of objectivity. As one member of an Ontario Royal Commission on asbestos told an EPA hearing in Boston in 1984, "Our governments, when they face a very complex issue and political problem, try to diffuse it a little bit by setting up commissions." The member, J. Fraser Mustard of the Canadian Institute of Advanced Research, explained further, "[I]t requires you to operate under a public inquiries act which I learned means that you have to take testimony under cross-examination.... [T]he people who appeared before you had to give their testimony under full cross-examination to not only the commission itself with its own legal counsel, but by all the parties of standing, labor, business and other groups."

The Ontario commission's finding was that "the [asbestos] risk in the average building was really negligible."

What was perhaps most significant in Mustard's comments was the testimony that immediately followed from a housewife in Maplewood, N.J. by the name of Susan Mazzochi "who knew of the dangers of asbestos only because of my work background.... [I]f I had not become involved in this issue in my community, no one would have."

Mazzochi identified herself as a former field researcher for Dr. Irving Selikoff, now retired from Mt. Sinai Hospital and the author of several studies EPA has relied upon in calling for a total ban on asbestos. She did not identify herself at the hearing—nor presumably to the principal of her local school—who she demanded shut down the building after she found "flaking pipe covers" with 40-50% chrysotile asbestos"—as the wife of Anthony Mazzochi, former occupational safety and health director of the Oil, Chemical and Atomic Workers International Union AFL-CIO.

Mazzochi was defeated twice for the presidency of the union after making comments such as: "Cancer is the cornerstone of the

American industrial process. There's an institutional imperative that says you have to kill people to produce things," he claimed. His wife, in turn, accused the school principals in her community of conducting "a campaign of lies and intimidation" against her and her family.

These are obviously matters of dispute.

What is not in dispute is the fact that Mazzochi is the "most prominent" among the "handful of labor leaders," in the words of Dr. Samuel S. Epstein in *The Politics of Cancer*, who have proclaimed we are in an era of "genetic confrontation," in which capital vs. labor confront each other across not just economic, but health lines as well. Mazzochi, in Epstein's words, sought out "a few independent professionals in the academic community" to assist him in pushing that confrontation forward. However, as Epstein further observed, "most of the focus of labor concerns on chemical exposures has been so far expressed in Washington, D.C. rather than at the grass roots level."

That may be because the people of Maplewood, N.J. have listened to the local public officials who have accused Mrs. Mazzochi and her husband of being radical, hysterical, angry, troublemakers, destroyers of property values. Mrs. Mazzochi says, "We have been accused of lying to the community and the community has been told to stay away from us."

It could also be that some of the Maplewood residents are readers of the presumably "radical" *New York Review of Books* where Susan Sontag, author of *Illness as Metaphor* has been known to write such revolutionary thoughts as: "[C]ancer is not just a disease ushered in by the Industrial Revolution (there was cancer in Arcadia) and certainly more than the sin of capitalism (within their more limited capacities, the Russians pollute worse than we do). The widespread current view of cancer as a disease of industrial civilization is as unsound scientifically as the right-wing fantasy of a 'world without cancer' (like a world without subversives). Both rest on the mistaken feeling that cancer is a distinctively 'modern' disease."

How can one reproach a man who has truth on his side?
—Robespierre

12 Cancer Games

What is distinctively modern about cancer is the fact that it has become big government. But unlike other big government agencies, it has not received as much critical attention as the Department of Defense, for example. One doctor quoted by James T. Patterson in *The Dread Disease*, a definitive history of well-intentioned but thus far ineffective anti-cancer initiatives, made the point clear in specific reference to the National Cancer Institute (NCI): "They subsidize and perpetuate ignorance in the mistaken belief they are benefiting humanity."

The accusation is bitter, but relevant—and cannot be ignored in any honest debate over EPA's proposed ban on asbestos.

Unfortunately, the debate, even with the benefit of the 1986 EPA cross-examination hearing remains tightly within the control of Washington bureaucrats. The hearing was completely within the hands of the agency. The chairman—or chief judge—Michael Winer, an EPA employee, was assisted by an EPA legal counsel, Kevin Lee. The other member of the adjudicating panel, David Dull, was a deputy director of the EPA office which would be responsible for enforcing an asbestos ban.

The witnesses were not required to take a formal oath to tell the whole truth and nothing but the truth, but were simply "reminded" they were subject to the False Statements in Government Act, which presumably makes it a crime to make a false statement in a government proceeding. No penalties were cited for infringement.

A formal protest was entered "as to the impartiality and fairness of a procedure which permits EPA counsel to stand as the chairman of this proceeding and have his colleague, Mr. Lee. . .be counsel for this proceeding."

Unlike royal commission proceedings in British common-wealth involving matters of public safety and health, no effort was made by the government to solicit a wide range of scientific and other opinion on which public policy might be framed. Nevertheless, attorneys for the affected interests, trade associations for asbestos producers—automatically suspected only of mercenary motives—were able to score significant points that forced the agency to commission new studies. The alternative would be to admit EPA might have been wrong in the first place.

The proposed ban, for practical purposes, has gone the way of the notorious OSHA generic policy of the late 70s, into regulatory limbo.

The first major point made in the hearings was that a retired employee of NCI, Dr. Marvin Schneiderman, was the sole outside scientist retained by EPA to review the evidence presumably justifying the ban. It also became evident in the proceedings that EPA had relieved exclusively on scientific studies of only one other scientist, Dr. Irving Selikoff, in formulating, not only its original asbestos-in-schools regulation, but the proposed ban.

Scheiderman, EPA's sole expert and a consultant to Clement Associates, was also one of the signatories of the by now notorious "Estimates Document," used to justify the abandoned OSHA effort to establish a "generic standard" for cancer regulation.

The primary source justifying EPA's proposed asbestos ban was a report prepared for the U.S. Consumer Product Safety Commission (CPSC) of which Schneiderman was one of seven authors. A CPSC ban on asbestos in consumer paints and putties had contributed to the crash of the Challenger shuttle. No consideration has been subsequently given by federal government agencies to the prophetic remarks of Barbara Hackman Franklin, a member of the CPSC, in a 1978 speech: "The consequences of whatever we do must be anticipated and alternative courses of action considered."

What did become apparent during the hearing was the EPA is 1) Very selective in what scientific advice it seeks; 2) The agency relies much more on in-house bureaucrats than internationally respected experts in making its decisions.

EPA's primary witness on the potency of asbestos was Bruce Sidwell, a plant biologist who has neither a medical degree, nor a doctorate in toxicology, epidemiology nor bio-statistics.

EPA's witnesses concerning the cumulative effects of asbestos were Amy Moll and Lynn Delpire. Both have bachelor degrees. Ms. Moll has a master's degree in public administration.

Under cross-examination, both admitted they were not qualified to measure asbestos fibers by any of the approved techniques; that they were not qualified to advise EPA regarding engineering and other controls; and, indeed, they had never been inside an asbestos product plant.

Cross-examination also reluctantly revealed that the EPA's estimated causes of death from asbestos originally placed at upwards of 1,800 over the next 12 years, were actually less than 1,000—by the agency's own figures then in 1986. Of course, by the time the agency announced the ban, the number of lives that would be presumably speared had been cut fivefold, down to 200.

● Moll and Delpire admitted 468 presumed cases of cancer caused by vinyl-asbestos flooring material, on re-examination of the evidence, drops down to "zero cases."

● Sidwell admitted EPA had overestimated the risk of persons contracting mesothelioma by 33%, reducing the estimate from 225 cases to 150.

● EPA conceded that its assumption that 50 cases of cancer would result from installation of asbestos cement pipe was 90 times higher than OHSA's. The more accurate figure would less than one case.

● Again, EPA conceded its estimate of cancer cases caused by installation of asbestos containing sheet material each year is 20 times higher than OSHA's estimate. That would be a little over one case a year.

● Further, and most significantly, EPA admitted its risk

assumption for full-time brake repair workers—the only potential danger to workers from asbestos currently and one critical for safety on the highways—was four times higher than OSHA's estimate, 400 rather than 100.

Sidwell admitted the mathematical model used by EPA rested on extrapolations from exposures 10,000 to 100,000 times lower than those faced by miners and industrial workers in the past. Sidwell further admitted on cross-examination that the scientific evidence "suggests" that chrysotile or "white" asbestos, whose fibers are less than five microns in length and are easily exhaled from the lungs are much less potent and, consequently, cause fewer cases of either asbestosis, mesothelioma or lung cancer.

Scheiderman, in his testimony, was not as forthcoming. . . . At one point, he was even admonished by Winer, the EPA chairman: "I'd like to caution Dr. Schneiderman in the future to be a little more responsive to the specific question asked."

Scheiderman, as one of the original seven member panel which drew up the CPSC study, said he reviewed the proposed EPA asbestos ban "only to determine whether it was a satisfactory statement of what I knew at the time was current knowledge. I wasn't asked to find whether it was consistent with anyone or anything."

The CPSC report was issued in 1983 after interested parties had an opportunity to nominate the experts on the committee, and drafts of its documents were released for comment and public hearings held. By contrast, Sidwell agreed that the initial support document justifying the 1982 asbestos-in-school regulation—which was savaged by outside reviewers—"relied upon the study by Dr. Selikoff of insulation workers."

"To a great extent, that's true," Sidwell said.

Under cross-examination, both Sidwell and Schneiderman conceded no effort had been made to solicit either the past or current views of the other six members of the CPSC committee on the advisability of the asbestos ban. Further, no attempt was made to incorporate—or even assess—the views of the Ontario Royal Commission, the British Health Executive study of Sir Richard Doll and Richard Peto, a National Academy of Sciences survey commissioned by EPA itself, or the dozens of published articles in the scientific literature which have established the potency of chrysotile or white asbestos is much less than other varieties, especially at levels considerably less than those experienced by the insulation workers studied by Selikoff.

In other words, the agency depended on 20-year-old studies as the sole source justifying a total ban on asbestos, which could have incalculable effects on the safety and health of millions of Americans. And the single scientific authority EPA paid to defend its position, Schneiderman, was a signatory of perhaps the most reviled document in 20th century scientific history, and one who had previously recanted earlier prophecies of a cancer epidemic—reluctantly.

A measure of Scheniderman's reluctance could be gauged by some of his answers under cross-examination.

Q. You were one of the co-authors of that [document] paper, is that right?

A. I was involved in its preparation.

Q. [I]t was given a good deal of publicity, primarily by the Secretary of Health, Education and Welfare at the time, Joe Califano?

A. Yes, Mr. Califano cited it in a press conference. I don't know whether you could characterize that as a good deal of publicity. . . .

Q. And part of the reason it got picked up by the press was that it attributes a very substantial number of cancers each year in this country to asbestos exposures in the past, isn't that correct?

A. You're asking me to look into the minds of newspaper reporters and I'm not an expert on reading newspaper reporters' minds. . . .

Q. And did this document get seriously criticized by a number of persons after it came out?

A. Oh, yes, oh, yes. It was very seriously criticized. . . .

Q. This is an excerpt from a document. . .[by] Richard Doll and Richard Peto [where] they say there: The speakers in the session devoted to asbestos came from a range of backgrounds, including Dr. Selikoff's department in New York, National Cancer Institute and various industries and universities, using some quite different epidemiological approaches. Several of these speakers devised numerical estimates of the proportion of U.S. cancer deaths currently due to asbestos, all of which were around one or two percent, rather than 13-18 percent. And no speaker or participant dissented from this consensus.

Q. Is this a correct description of the conclusion of this broad expert group when they reviewed the Estimates Paper three years after it came out?

A. There were extensive discussions and presentations of material relating to asbestos, and the potential number of deaths. The group, the various people who reviewed this, indicated that they felt the estimates that appeared in the so-called Estimates Paper, which by the way were consistent with some estimates made by Professor Selikoff, were much too high.

And were of the order, the one set of estimates were too high. As I said, there were two sets of estimates. The estimates at the meeting [Banbury meeting, Cold Spring Harbor Laboratory, N.Y.] ranged from about 5,000 to about 12,000, as I recall. In contrast with the higher of the two estimates that appears in this paper which were about 50,000 to 60,000 which were consistent with some testimony that Professor Selikoff gave to a Congressional committee at one time. The second estimate of what's in this particular paper is about 13,000 deaths, which would be within the range of the estimates given by the people at the Banbury conference, probably at the—certainly at the high end of the range. . . .

Q. But certainly the estimates as of 1981—and I guess I hear you correctly—were substantially lower—than the high estimate in the Estimates Paper?

A. They were substantially lower than the high estimate in the Estimates Paper. . . .

Q. [I]t's the newspapers who. . .grabbed the higher number, which I believe was clearly included in the Estimates Paper, is that right?

A. Oh, yes, the high number is there.

Q. So then as I understand it, the estimate of the fraction of cancers in this country, at least the high estimate in the Estimates Paper, was an estimate that's been reduced in the process of scientific review by a broad variety of people, including, as you've indicated, Dr. Selikoff and his people, it's an estimate that's been reduced by an order of magnitude over the years.

A. Well, let's see, the estimate was about 50,000 or 60,000 in this paper, compared to about 12,000, but it's quite an order of magnitude but certainly been reduced. . .substantially reduced. You know, I'm not going to argue with it about that particular point.

Q. And just looking again at the discussion of the numbers by Professors Doll and Peto, they talk about it being reduced from 13 to 18 percent (total cancer cases due to industrial, primarily asbestos exposure) to one or two percent, which is why I got the impression it's about an order of magnitude.

A. Well, as you know, Richard Peto and I were joint editors of that Banbury volume so we were there at the meeting, and Richard kept—Richard Peto kept insisting on using the lowest estimates that people had estimated of these numbers of deaths. And that's what leads to this. He probably used about 5,000 rather than 12,000, perhaps 12,000 or 13,000 or even 8,000, so I think they've overestimated the overestimation.

Or he's underestimated the number of cases, so I think there is some leeway. What do you have, you have a swing of the pendulum, he's counteracting or arguing against what has been said, so he swings too far in the other direction.

Q. Either way there has been a substantial reduction, though?

A. Either way, there has been a substantial reduction."

Everett Dirksen, the late Republican leader of the Senate, had a way of describing this kind of numbers game. "A billion here, a billion there, after a while you get into real money," he said.

The question here is when does 8,000, 12,000, 13,000, 50,000, 60,000 lives sacrificed to—or conversely when does 1,800, 1,000 or as few as four-lives-a-year saved from asbestos—become worth spending $3.1 billion to $15 billion in schools and $60 to $200 billion in all the buildings in the country that might contain asbestos.

There are five issues, more properly questions here, 1) scientific, 2) political, 3) legal, 4) practical, and 5) cultural.

The first was addressed by C. P. Snow, the scientist and novelist famous for his comment about the "two cultures" of science, and, for the lack of a better phrase, "the rest of us." In his book, *The Search*, he talks about the hard necessity of self-discipline among scientists in writing about a decision to expel a promising young scientist from a university post for a relatively minor infraction:

"[Y]ou see we both committed a crime against the truth. A crime in good faith, admittedly, honest, simply a mistake. Your mistake, if I may say so, was even stupider than mine. But there we were: we issued false statements. Now if false statements are to be allowed, if they are not to be discouraged by every means we have, science will lose its one virtue: truth. The only ethical principle that has made science possible is that truth shall be told all the time. If we not penalize false statements in error, we open the way, don't you see, for false statements by intention. And, of course, a false statement of fact, made deliberately, is the most serious crime a scientist can commit. There are such, we both know, but they're few. As competition gets keener, possibly they will become more common. Unless that is stopped, science will lose a great deal. And so it seems to me that false statements, whatever the reason, must be punished severely as possible."

It is obvious Snow was a better prophet than judge and the scientific community has lost any real capacity for self-discipline. Society, as a whole, is suffering as a consequence. But scientists, unlike doctors or lawyers, are not licensed or likely to be. Consequently, some other method is needed to determine whether they are making false statements, whether deliberately or in good faith.

This consideration leads to perhaps the most important of the recommendations or suggestions with which this book concludes. It does not deal directly with the asbestos situation, but it will hopefully prevent such cases from occurring in the future by addressing the political, legal, practical and cultural realities.

It calls for the passage of what can be called a "Public Inquiries Act," governing the development and enactment of all regulations affecting public safety and health. The regulation would apply to EPA, CPSC, OSHA and all other agencies with authority to promulgate regulations affecting public welfare.

The law would not, in any way, weaken their authority to act vigorously on behalf on the public. But it would take them out of the ludicrous situation of being simultaneously the prosecutor, judge, jury and enforcer of their own regulations.

In future, under the provisions of the act, they will be confined to the role of prosecutor and enforcer if the regulations are enacted. When and if the agencies believe a new regulation is needed or an older one revised, a public notice would be published in *The Federal Register*, as is currently the case, but with one significant exception. The notice would also call for the summoning of a public inquiries commission to hear evidence under oath and with full examination of witnesses, under summons if need be, of the facts of the matter.

The commission and its staff would be entirely independent of the agency and would go out of business upon rendering its verdict. The composition of such a committee is open to discussion, but, as a practical matter, it would be advisable to base it on the well-established tradition of British Commonwealth Royal Commissions being composed of three members. In deference to the agencies, the chairman could be their nominee. The two other members, one representing citizen groups such as the Environmental Defense Fund, and the other, scientific bodies such as the National Academy of Sciences, could be nominated by those groups or university presidents.

The choice of the judges should be the responsibility of the President in line with the precedent established by President Theodore Roosevelt in establishing the Remsen Commission which adjudicated the first attempt to ban saccharin back in the early 20th century. Public health should be as much a political issue as war and peace, for it is the primary responsibility of government to protect the lives of its citizens from all sources of danger.

As an alternative, the final choice of the members would be made by the House and Senate chairman and ranking member of the appropriations committee for the agency. Responsibility would thus be fixed and clearly understood. All hearings of such commissions would be held publicly, of course, but also in areas and communities, which in the judgment of the commission, would be most directly affected by the regulations—proposed or actual waste site dumps, for example.

The agencies, EPA, OSHA, CPSC, etc., should be free to advocate their proposed policies through public information strategies. But they should always be clear and explicit that the message is what the agency recommends, but not what is authorized to enforce unless given clear and explicit authority to do so.

Second, due consideration should be given to the elevation of EPA to cabinet status, as has been recommended by a number of senators. The severe criticism this book has directed at the agency underscores rather than minimizes its importance in the lives of all Americans.

For EPA needs a charter based on law and science, rather than politics.

It can no longer be the bastard child of a self-serving Presidential executive order which created it, or a series of equally self-serving laws subsequently passed by Congress.

A series of public hearings by both the House and Senate are needed to draft a basic law defining the role and purpose of EPA and codifying the mishmash of laws under which it operates. Ironically enough, William Ruckelshaus, the EPA administrator under whom some of the agency's worst abuses occurred, would probably be the most eloquent witness in support of such a charter.

He knows from experience.

Two incidents from the history of OSHA further illustrate the absolute necessity of strict White House control of environmental regulation and rigorous Congressional oversight under both political parties.

When OSHA was created, the law gave the Labor Department a year to scrutinize and closely examine the consensus standards for safety and health developed by industry, labor and insurance companies over the years. Instead, all the standards were given the force of law by publication in The Federal Register within 30 days after the enactment of the law.

Why?

Bureaucrats don't build empires—and get promotions—by hiring a few dozen analysts. They do so by hiring hundreds of inspectors to enforce regulations, which, in this case, bypassed and ignored the standards developed by the Conference of Governmental Industrial Hygienists governing the control of then known carcinogens. Those standards were inadvertently—or deliberately dropped—from *The Federal Register* notice establishing OSHA's regulations. The absence of the Conference of Governmental Industrial Hygienists' implicit acceptance of the concept of threshold levels led to the attempt to set a generic standard of carcinogens and the Estimates Document based on a premise that asbestos statistics could be used to justify risks from carcinogens down to zero level.

This absurdity was underscored when Thorne Achuter, then assistant director of Labor for OSHA, was hailed before a Congressional committee for failure to issue an "emergency" temporary standard on asbestos. He was cross-examined, in the summer of 1983 during the period when Anne Gorsuch Burford was about to be ousted as director of EPA. Her panicky evacuation of Times Beach, Missouri, because of a presumed danger from dioxin, the contaminant in Agent Orange, did not save her from Congressional wrath. Nevertheless, Auchter was determined not to share her fate.

And Auchter had little reason to expect any mercy from Rep. Barney Frank (D. - Mass.). Frank, at the time was the chairman of a subcommittee investigation of an alleged "Failure to Regulate—Asbestos: A Lethal Legacy:"

Q. What would be the harm of having a temporary emergency standard. . . .?

A. We don't believe it would be enforceable. . . . An emergency temporary standard violates the public's right to be involved in the rule-making process. As we have demonstrated earlier. . .the regulation of asbestos. . .is a very complicated and difficult issue."

A few days after this exchange, Auchter had a change of mind after meeting "a recognized expert in the epidemiology field"—none other than Irving Selikoff. Auchter's subsequent issuance of an "emergency" temporary regulation was "precipitated," the U.S. Court of Appeals for the 5th Circuit in New Orleans later declared by the meeting with Selikoff. The court made the observation in striking down the "emergency" regulation. Auchter had resigned his position the day before the decision was rendered. Unlike Burford, who forced people out of their homes in Times Beach, Missouri, because of a groundless fear of dioxin in a vain effort to keep her job, Auchter, at least, had not been forced out of his job.

But we haven't learned from experience. Neither Auchter, Burford, Congress, the media nor the general public have been the beneficiary of fully understanding what happens when under rigorous cross-examination, the potency theory—no safe threshold level, one asbestos fiber can kill—the central premise of the environmental movement is closely looked at.

It becomes, in the words of a New York Times editorial "Orangemail," a racket.

The Times in the "Orangemail" editorial on May 9, 1986 commented that a settlement worked out by U.S. Judge Jack Weinstein in cases involving exposure to the defoliant, Agent Orange, used extensively in Vietnam was a "far better solution than the chaos of the asbestos case, where the manufacturers have gone bankrupt and the victims still await compensation while the lawyers squabble over fees. The settlement achieved by Weinstein may prove a model for similar cases."

The Agent Orange case was inspired, in large measure, by

another set of *New Yorker* articles like those written by Paul Brodeur on asbestos. The Agent Orange series was written by Thomas Whiteside and subsequently published under the title of *The Pendulum and the Toxic Cloud* in 1979. A year later, a TV documentary reflecting its views—and the pervasive guilt most Americans were feeling about the shabby treatment accorded Vietnam veterans was produced by Bill Kurtis for CBS News.

But it was apparent even by that time to scientists that the dangerous element in Agent Orange, dioxin, was a contaminant produced by a natural process of combustion. An article in *Scientific American* concluded dioxin has "now been found in the effluent and ash of so many combustion products that there is no longer any serious argument about their formation in combustion, although the precise nature of the process remains obscure."

Dioxin or TCDD is, in other words, is a "ubiquitous chemical, particularly in industrial nations," the article concluded. Therefore, it cannot be logically held accountable for disease caused allegedly by Agent Orange alone, as Judge Weinstein ruled.

But why did the case arise in the first place, and who was ultimately responsible. The *Times* placed the responsibility squarely where it belonged: 1) the federal government which in World War II ordered the use of asbestos—without any controls—in warships and the similar use of Agent Orange in Vietnam; 2) the lawyers who exploited the fears and resentments of veterans who had to settle for an average settlement of $8,000 while the attorneys walked away with hundreds of thousands in legal fees

"Devastating crops and forests in Vietnam was the government's idea," as the *Times* editorial pointed out, "and the government's moral duty to the veterans transcends that of the chemical companies that supplied the herbicide. Whether or not Agent Orange is a real cause of disease, it has become a metaphor for the veterans' sufferings."

The question still remains, in the words of another *Times* editorial, this one of August 13, 1986, "What was the truth about Agent Orange?"—or any carcinogen, such as asbestos, whose potential or real danger is in question.

"The herbicide was sprayed for 10 years over forests and roadsides in Vietnam to reduce the enemy's cover and prevent ambushes," the editorial observed. "Whether justifiable or not, the spraying probably saved the lives of many soldiers"—just as asbestos saved the lives of many sailors, soldiers and merchant seamen during World War II.

"But the herbicide was contaminated with minute amounts of an impurity, dioxin, which the survivors cite as a cause of their present illneses," the editorial continued. "Dioxin is a fearsome poison. But poisoning also depends on dose. The dioxin in all the herbicides spread during 10 years of war, over six million acres, amounted to just 368 pounds. Most of that stayed within the thick forest canopy, with just five percent reaching the ground.

"Could soldiers in a sprayed forest, or even soldiers sprayed directly by accident, have picked up a significant dose of dioxin? It's far from likely. Could hazardous amounts of undergraded dioxin have accumulated in the soldiers environment? Conceivably, but no way probable.

"Complaints about Agent Orange didn't begin until six years after spraying had ceased. Cancer apart, few known diseases have such a latency. And the diseases and birth defects the veterans complain of are widespread throughout the population.

"The pilots who sprayed the herbicide were exposed to it daily. In their tour of duty they received a thousand times more than ground troops being sprayed directly. Yet a survey of the pilots' health completed last February showed no unusual incidence of diseases, including the three dioxin is though most likely to cause.

"Other studies still underway may reveal some subtle pattern of ill health. But so far there's no reason to suppose either that the veterans were exposed to significant amounts of dioxin in Vietnam, or that they have any symptoms for which dioxin might be the explanation.

"How on earth," the editorial asked, "could such a tenuous case have ended in a $180 million settlement?"

And how on earth could such an less tenuous proposal such as EPA's ban on asbestos in buildings end up saddling the country with $160 to $200 billion of wasted money, and the wasted lives of thousands of asbestos removal workers?

Well, for one reason, the lead attorney in the Agent Orange case was Victor Yannacone, who along with Dr. Samuel Epstein and Dr. Arthur Upton founded the Environmental Defense Fund back in the late 60s to "Sue the Bastards" and ban DDT. And if DDT hadn't been banned the *Times* of that time would have been mighty miffed.

But the more fundamental answer is that we, as a people, persist in seeing cancer as a metaphor for everything else that ails or disturbs us, and not as a disease or a set of diseases, that may never be curable, any more than heart disease is. We have a right to the pursuit of happiness—but that doesn't bring with it a guarantee of immortality.

Government, even in its best state, is but a necessary evil;
in its worst state, an intolerable one.

–Thomas Paine

13 Uncle Sam Flunks

Like it or not, we're all going to the graveyard, just as sure as all old cars end up in the junkyard. The only question is which goes first, the engine–the heart–or the chassis–the body. People rust out just like cars, but we call it cancer.

There may well be a cure for cancer, or, at least the pain and suffering it imposes before a merciful death, but we might be well advised to consider the words of Susan Sontag, writing in *Illness as Metaphor* of the dangers of cancer becoming a paranoid vision.

> [O]ur views about cancer, and the metaphors we have imposed upon it, are so much a vehicle for the large insufficiencies of this culture, for our shallow attitude toward death, for our anxieties abut feeling, for our reckless improvident responses to our 'real problems of growth,' for our inability to construct an advanced industrial society which properly regulates consumption, and for our justified fears of the increasingly violent course of history. The cancer metaphor will be made obsolete, I would predict, long before the problems it has reflected so persuasively will be resolved.

In trying to think about and accept that, the words of Judge Jack Weinstein in refusing to accept the "expert" testimony of Dr. Samuel Epstein, author of *The Politics of Cancer*, the bible of the environmental movement, in the Agent Orange case are instructive.

Weinstein, in his final ruling said he would not have allowed either Epstein or the other doctor, Dr. Parry M. Singer, representing the plaintiffs, to testify as expert witnesses. They were, in his words, "legally incompetent."

That ruling, unfortunately like so many others that impeach the case of the environmental movement and EPA, OSHA, CPSC, etc., was never reported in the news pages of the *New York Times* or any other paper or television network. But Weinstein's comments are worthy of reporting.

His assessment of Singer's arguments were his "conclusionary allegations lack foundation in fact. Singer's analysis, in addition to be speculative, is so guarded as to be worthless," the judge ruled.

The judge's assessment of Epstein's testimony was, if possible, even more withering:

"An extensive deposition of Dr. Epstein dated April 11, 1985 adds nothing of a substantive nature to the affidavit, but consists of a devastatingly successful showing of his lack of knowledge of the medical and other backgrounds of those on whose behalf he submitted affidavits."

Weinstein's opinion continues:

"On the issue of cancer, Dr. Epstein himself has acknowledged that the approach he uses in Agent Orange of ignoring all other possible causes of cancer is invalid. . . .

"The point is that Doctors Singer and Epstein rely on hearsay checklists to garner essential facts about plaintiffs and on inapposite literature to reach their conclusions. They ignore more relevant studies and fail to show how the myriad illnesses at issue are more likely to have been caused by Agent Orange than anything else."

The folk singer Joe Jackson had put it in simpler terms a couple of years before. If everything causes cancer, can one thing be blamed?:

> Everything, everything gives you cancer.
> Everything, everything gives you cancer.
> There's no cure, there's no answer.
> Everything gives you cancer.

Weinstein's opinion went further:

"His deposition also shows that Dr. Epstein's conclusions are based 1) on incomplete or complete lack of knowledge of the family histories of the plaintiffs; 2) complete lack of knowledge concerning each plaintiff's occupational exposure; and 3) incomplete or complete lack of knowledge concerning the likelihood of exposure to Agent Orange in Vietnam.

"[T]he speculation and unfounded assumptions underlying [the] testimony of [Singer and Epstein] decrease the probative value, perhaps to the level of the gossamer.

"There is a strong probability that the doctors' testimony would mislead and confuse at least part of the jury. Establishing the low probative value of the affidavits would entail an unwarranted expenditure of time and effort.

"In sum, the court finds the expert evidence of Doctors Singer and Epstein inadmissible."

Had the case gone to trial, the judge ruled, "a directed verdict on the close of the plaintiffs' case would have be required" on the behalf of the defending companies.

The Agent Orange case has been settled.

Asbestos cases, however, can be expected to continue forever in an endless cycle. What Dr. Robert N. Sawyer, the leading authority of asbestos control has called "the third wave" of suits is already cresting as new cases are being filed as workers are being exposed to excessive amounts asbestos because of poorly managed abatement projects.

Congress has, with great reluctance, already recognized this by the passage of some amendments to the Asbestos Hazard Emergency Response Act (AHERA) permitting extensions of time for the development of management plans for asbestos present in school buildings. But there is a growing recognition, in the words of Michael Gough, currently biology director of the Congressional Office of Technology Assessment, that much more has to be done to correct the past errors of the federal government in both asbestos and other carcinogen control. That is reflected in the title of his article: "Uncle Sam Flunks Asbestos Control in Schools," in the Spring, 1988 issue of

Issues in Science and Technology. Its subtitle is: "The EPA's effort to protect students from breathing cancer-causing asbestos fibers could actually increase their risk."

Gough pointed out the new EPA rule while "addressing an admittedly difficult problem has several fundamental shortcomings." It provides no rigorous standards for determining when asbestos-containing materials pose little or no risk and can be safely left alone. The agency merely shifts the burden. As EPA admits in its description of the rule: "This process is based on the responsibility of local officials, with input from specially trained experts, to develop management plans to implement appropriate measures that will abate the risk of asbestos depending on local circumstances."

But the experts aren't there as Congress has tacitly admitted by the passage of the deadline extension amendment. And the quality of those turned out is indicated by the derisive comment attached to them by Rep. Thomas Luken (D - N.J.), House sponsor of the amendments, "five day wonders." The products of these programs, which cost between $350 to $475 for three day courses and $600 to $900 for five days, have included high-school dropouts and illegal aliens who spoke little English, according to an article in *Engineering News Record*, "Asbestos: Hazard and Hysteria," in the June 2, 1988 issue.

The only certification needed is a 70 percent score on a simple multiple choice test.

Another major deficiency in the EPA rule, Gough points out, is that it "also ignores numerous scientific studies indicating that asbestos airborne concentrations are, in fact, considerably lower than the levels based on measurements made by agency contractors several years ago. In fact, EPA cut its earlier exposure estimates 10-fold—to an average concentration of about 10 nanograms per cubic meter in rooms with asbestos containing materials—but even these levels are far higher than what other surveys suggest."

Gough cited, as another example, that researchers at the National Institute of Environmental Health Sciences have reported that "the average concentrations in the vast majority of buildings with asbestos as surfacing material do not differ significantly from background."

"It is not possible to differentiate indoor air samples with regard to asbestos fiber content," Dr. Morton Corn of Johns Hopkins University and former assistant secretary of Labor for OSHA, has said.

Corn pulled together data from "hundreds of samples in buildings with water and abusive damage to the asbestos containing surfaces." In other words, the material had been damaged.

Even then, this is what the samples showed:

Sample Set	No. of Samples	Median Fiber Concentration
Air of 48 U.S. Cities	187	0.0005 fibers/cc
Air in U.S schoolrooms without asbestos	31	0.00054 fibers/cc
Air in Paris buildings with asbestos surfaces	135	0.00006 fibers/cc
Air in U.S. buildings cementitious asbestos	28	0.00026 fibers/cc
Air in U.S buildings with friable asbestos	54	0.00064 fibers/cc

Corn pointed out the clearance level EPA has set as the measure for completion of control or removal projects is 0.01 fibers per cubic centimeter of air (f/cc), thousands of times higher than average back background levels. "You'd have to seal up a building for six months or more after a removal project and run the ventilating system full to get the levels down to what they were originally," Dr. Malcolm Ross of the U.S. Geological Survey has estimated.

Corn estimated removal is only warranted in 10-15% of the buildings where it is found. He has incorporated that finding in manuals he has written for the U.S. General Accounting Office, the Library of Congress, the U.S. General Services Administration, which manages most federal civilian buildings, and International Business Machines, Inc. It could be argued the federal government has one rule for itself and another, far more stringent one, for local communities.

Yet EPA continues to resist air monitoring, despite the fact that in its own study of federal buildings, showed as Gough reported: "For buildings in which buildings were judged to be in good repair, the media airborne concentration turned out to be 3.5 nanograms per cubic meter. For buildings with material in poor repair, the median concentration was only 0.25 nanograms," That Gough pointed out is *"just the opposite of what inspections might have logically predicted."* And the inspectors who came up with those findings were trained according to EPA standards.

If air monitoring is essential in determining whether the asbestos in buildings is a problem, it is indispensable in asbestos control projects. A study done by Battelle Laboratories, perhaps the best testing laboratory in the country, showed such removal projects often release great clouds of asbestos fibers, even when performed under close EPA supervision.

"At a minimum," Gough concluded, "EPA should allow school boards to monitor air quality in areas identified by inspectors as containing friable asbestos that is significantly damaged. Now, the schools are obligated to take immediate action—which, given the EPA's lack of guidance about best-suited remedial strategies, will often mean removal."

"Knowing when *not* to act can be as important as knowing when to act," as Gough observed.

But that won't be easily done, as the history I have recorded here illustrates. EPA has consistently dodged responsibility for assessing the dangers associated with asbestos and developing effective means of dealing with them.

EPA can never be persuaded; it must always be pushed.

14 A Call to Action

A broad-based coalition of parents, taxpayers, home and building owners, realtors, even contractors—the entire voting community—must be formed to deal with a situation that is getting completely out of control. It can be an adjunct of existing organizations or entirely independent.

But it must be formed now or it will too late—because no one else can represent the interests of the owners, residents and occupants of homes, offices, factories and other buildings affected by EPA's asbestos regulations.

The only alternative is to wait for EPA—and asbestos hysteria—to come knocking on the door.

This coalition should have two immediate goals, a third and fourth intermediate one, and a fifth long range.

The first is to seek an amendment to AHERA that would require air monitoring be the means of determining whether an asbestos problem exists. It would apply to all buildings, public and private, and require records be kept and be construed as legal defense against action taken—or not taken—in buildings. The cost of any other alternative will be many times greater.

The second is an amendment to AHERA that would give consensus specifications developed by the National Institute of Building (NIBS) the force of law. NIBS, which is based in Washington, is a semi-official arm of the federal government. It has a Congressional charter and derives one quarter of its income from a trust fund established by Congress.

The law setting NIBS up, the Housing and Community

Development Act of 1974, provided that: "Every department, agency and establishment of the Federal Government having responsibility for building or construction-related programs is authorized and encouraged to request authorizations and appropriations for grants to the Institute for its general support, and is authorized to contract with. . .the Institute for specific services."

And EPA did just that in participating in the development of the NIBS specifications under the guidance of an 85-member task force composed of representatives of every major architectural, engineering, building and repair association.   The specifications are regularly updated to reflect current knowledge, and are keyed to the MASTERSPEC system for writing building contracts produced by the American Association of Architects.

Without a uniform set of standards for dealing with asbestos, chaos will result. In New York City, for example, a local law applied to every building, with the sole exception of one-family houses, requires an inspection for asbestos before a nail can be driven in a wall. These NIBS consensus regulations are not only needed, there is ample precedent for them. That was the way safety and health regulations were developed by the Labor Department before OSHA came along and created the current chaotic situation where judges, lawyers and politicians agree on what's feasible, not engineers, architects, contractors and building owners.

The third and fourth recommendations could be carried in tandem. Passage of a "Public Inquiries Act," setting up a modified version of the royal commission system for dealing with environmental issues takes precedence. The key elements in such a law are 1) a panel of commissioners independent of the agency which proposes and would implement new or revised regulations; 2) Full cross-examination of witnesses—under oath.

The fifth recommendation is the passage of an EPA charter, a single law spelling out the agency's mission and making the administrator—or secretary if the agency were converted to a cabinet department—responsible to one authorizing committee in the House and Senate. Until that happens, until the agency or department no longer has to play to the political agendas of dozens of senators and hundreds of representatives scattered among more than 82 committees and sub-committees, it will continue, in the name of the environment, to destroy it.

That's going to take time and a lot of public education. But the framework for understanding has been solidly laid by two leading scientists, physical and political. The first is Dr. Bruce Ames, chairman of the department of biochemistry and the University of California at Berkeley, the physical scientist. The second is Dr. Aaron Wildavsky, also at Berkeley, former president of the American Political Science Association, the political scientist.

It was Ames, the biochemist, who invented the simple test for identifying mutagens, gene-changing agents, that has fundamentally changed our understanding of carcinogenesis. Life itself "causes" cancer, like it or not.

"Far from being a rarity," Ames has said, carcinogens are "literally everywhere. . . . Most food, including everything broiled or roasted, is loaded with carcinogens."

Carcinogens abound naturally in such foods as uncooked corn, nuts, peanut butter, bread, cheese, fruit, beets, celery, lettuce, spinach, radishes, mushrooms and rhubarb.

"Cooking our food also generates mutates and carcinogens, as all browned and burned material contains them," Ames wrote in a 1983 article in *Science,* published by the American Association for the Advancement of Science.

"It takes a Los Angeles resident one year of breathing smog to equal the burned material that a heavy smoker inhales in one day," his article continues, "And we eat even more browned food than the heavy smoker inhales, though we don't know the risk from this."

Ames' credentials are enhanced by the fact he was once a champion of the environmental movement. He helped persuade CPSC to ban Tris, a flame retardant used in children's pajamas when it was linked to cancer.

He was also an early advocate of limiting workers' exposure to ethylene dibromide (EDB), a soil fumigant banned by EPA in 1985 because it caused cancer in rats. But he urged the agency, however, not to outlaw EDB unless it recommended an equally effective substitute to help farmers fight crop pests–the kind of risk-benefit question also raised about asbestos.

"People hear carcinogen and they get so excited," he said. "But you have to think of amounts. Sunlight is a carcinogen, but you don't tell your friend not to walk across the street in the sun."

Wildavsky, in a 1982 book called *Risk and Culture*, written with Mary Douglas, has provided the sociological compass points: "There used to be an accepted difference between primitive ways of thought and our ways of thoughts. . . . [I]t was summed up in the primitive attitude toward danger and death. . . . After millennia of our human past in which dangers were said to be caused by witchcraft and taboo breaking, our distinctive achievement was to invent the idea of natural death and actually believe in it."

Environmentalism has pulled us back into "the primitive worldview [where]. . .everything that seems abnormal is explained by the intervention of mysterious agencies called into existence by common fears and common perceptions, the mystic participations of the culture. . . .

"[T]he defining feature of primitive mentality was to try to nail a cause for every misfortune, and the defining feature of modernity, to forbear to ask. . . . [Now] moderns using advanced technology are asking those famous primitive questions as if there were no such thing as natural death, no purely physical facts, no regular accident rates, no normal incidence of disease. We seemed to have changed places–no, we have joined the primitives. They demand an autopsy for every death; the day we do that the essential difference between our mentality and theirs will be abolished."

In a 1988 book, *Searching for Safety*, Wildavsky argues safety is only found by taking risks. As an example, no one died in the space program, until a totally irrelevant fear about asbestos in hairdryers led, with all the inevitably of a Greek tragedy, to the crash of the Challenger.

Wildavsky's thesis and the intellectual framework he laid

down is best summarized in a review of his book in *The Wall Street Journal* of May 19, 1988 by Edith Efron, author of *The Apocalyptics*.

Efron wrote:

"In the past two centuries, those competitive capitalist societies that have systematically courted technological risk have provided their populations with steadily increasing and historically unprecedented safety, health and longevity. Most of this 200-year-old phenomenon has been unplanned and unintended, and relatively little of it has been due to medical advances. It was only 45 years ago that genuinely scientific medicine was born, with the arrival of antibiotics."

Safety and health, says Wildavsky, are primarily a function of wealth. And concern about safety and health is being expressed in an inverse ratio to real, as opposed to perceived dangers.

"The contemporary notion, espoused by the risk-adversaries, that life is 'being sacrificed for material' gain thus violates two centuries of evidence," according to Wildavsky. "Risk adversaries assume that danger can be outlawed. They further assume that safety and danger can be severed from one another."

In fact, as Wildavsky demonstrates, safety and danger are the very warp and woof of both nature and technology. He points out: "To focus exclusively on sources of danger is to direct thought and resources toward an infinity of hypothetical and acutely low-probability risks (i.e. one molecule can cause cancer) while ignoring—and damaging—the inexorable related sources of safety."

In medieval Europe, totally encompassing systems of philosophy, exemplified by St. Thomas Aquinas in the Christian tradition and Maimonides in the Jewish seemed to offer answers to all the questions of life and death. New knowledge was apparently not necessary. It was only necessary to study the masters and acquire their knowledge. The result was scholars became pendants. But bored by just restudying and re-writing the old masters, many became fascinated by the new system of counting introduced from Islam, Arabic numbers.

The numbers were infinitely more subtle than the old cumbersome Roman numeral system. They could be easily added, multiplied, divided, subtracted, and raised to the Nth power—but they had no practical uses. The price of a loaf of bread or a street designation could be just as easily enumerated using the Roman system. The Arabic numbers were just toys, and soon the scholars who played with them were mocked for "debating how many angels could dance on the head of a pin."

But then the anonymous geniuses of the late Middle Ages, the builders of the enormous cathedrals which were slowly built over the centuries all over Europe, found they had a very practical use: measurement. They could not only be used to measure length and depth, but breadth and height, the weight of an arch, the geometry of a buttress, the glory of Chartes.

That was the paradigm of what was to happen with the emergence of capitalism five centuries later. The classic trial and error learning which created Chartes created the machine and a good life that could only be prayed for in centuries past. Numbers became tools and wealth—and health—were created by millions of anonymous builders in capitalist countries. "Decentralized, unplanned and propelled by competitive groups, trial-and-error learning is a high-speed flexible process consisting of hypothesis testing, feedback, discovery of error (danger), and incremental correction," Efron wrote in reviewing Wildavsky's book, "It discovers and controls or arrests danger as it emerges."

Learning is not a slave to theory; it is the creator of new theories.

"Only an ideologue knows the truth; the scientist knows only a theory."

Those words of Dr. Sherwin B. Nuland, professor of surgery and scientific history at Yale University in his book, *Doctors: The Biography of Medicine*, are my theme in this book.

Carcinogens dancing on the cursors of computers are the angels of death dancing on the pinhead of our time. They are slaves of a theory as obsolete as medieval scholasticism. Yes, we are "living in a sea of carcinogens," as Rachel Carson wrote in *Silent Spring*.

But there is more in the sea than sharks or barracuda—or

crabs. There are algae, minnows, bluefish, shrimp, seaweed—or silent stingrays. If life is to continue and improve, what has been called "the search for a vanishing zero" or if-you-can-detect-it-ban-it theory is obsolete. Computers and other measurement agents have to be used to measure not only the length and width, but the breadth and height of our world, the weight of knowledge, the geometry of insight.

The central environmental thesis that everything in the world is interrelated is correct. But it can no longer explain reality in presumably simple cause and effect relationships. These synergistic interrelationships, exemplified by cigarettes and asbestos, constitute the real danger, but have not, as yet, been adequately explored.

And exploration brings no guarantees, either political or real.

"Shortly after taking over as the new administrator of the Environmental Protection Agency in March, 1983, William Ruckelshaus addressed a meeting of the National Academy of Science in Washington," David Dickson wrote in *The New Politics of Science:* "Ruckelshaus succeeded Anne Gorsuch. . .who had resigned after escalating public and congressional criticism of her management of the agency made her a political liability of the Reagan administration.

"The first task therefore facing Ruckelhaus was to raise the internal morale and, even more important, the external credibility of the agency. Both, he suggested to the scientists gathered at the NAS, could be achieved by giving greater weight to the importance of science in regulatory decisions.

"Ruckelshaus told the scientists that the EPA was not going to emerge from the disorganized situation in which it had been left by Burford without a significantly higher level of public confidence in its activities."

"The polls show us that scientists have more credibility than lawyers or businessmen or politicians," Dickson quoted Ruckelshaus as quipping, "And I'm all three of these."

"I need your help," Ruckelshaus concluded.

It seems fair to say the rest of us need the help of scientists more.

But if the public does not help itself by demanding effective action from Congress, based on scientific facts rather than emotional posturing, it will only have itself to blame for AHERA and other misguided laws.

Ultimately, we in a democratic society are responsible: The animals, the fish, and the plants do not vote.

We do.

The asbestos issue is important in itself, but it is more important as a paradigm—or a parable—about the way people perceive, learn about, and try to deal with life and death.

I do not know the answers. But I do know life is real, and death inevitable—but in that brief time between birth and the grave, we are the only creatures capable of believing—as Socrates said, the "only life worth living" is self-examined.

Science is the road.

Faith and poetry are the roadmaps.

FAITH:

To everything there is a season, and a time to every purpose under the heavens:

There is a time to be born, and a time to die; a time to plant and a time to pluck up that which has been planted.

A time to kill, and a time to heal; a time to break down and time to build up:

A time to weep, and a time to laugh; a time to mourn and a time to dance.

A time to cast away stones, and a time to gather stones together:

A time to embrace, and a time to refrain from embracing:

A time to get and a time to lose; a time to keep and a time to cast away:

A time to rend and a time to sew; a time to keep silence and a time to speak,

A time to love and a time to hate; a time of war and a time of peace.

—Ecclesiastes

POETRY

We are the music makers
   And we are the dreamers of dreams,
Wandering by lone sea-breakers,
   And sitting by desolate streams;–
World-losers and world-forsakers,
   On whom the pale moon gleams:
Yet we are the movers and shakers
   Of the world for ever, it seems.

With wonderful deathless ditties
   We build up the world's great cities,
And out of fabulous story
   We fashion an empire's glory:
One man with a dream, at pleasure,
   Shall go forth and conquer a crown;
And three with a new song's measure
   Can trample a kingdom down.

We in the ages lying
   In the buried past of our earth,
Built Niveveh with our sighing,
   And Babel itself with our mirth;
And o'erthrew them with prophesying,
   To the old of the new world's worth;
For each age is a dream that is dying,
   Or one that is coming to birth.
                    –Arthur O'Shaughnessy (1844-1881)

We may be God's steward on earth. But we are not God–and asbestos is a parable. We know how to ask the questions.

We can only guess at the answers. But we must try, or reject the one thing that makes us human... our minds.

> *The political and commercial morals of the United States are not mere food for laughter, they are an entire banquet.*
> —Mark Twain

A Chronology

Pre-Christian Era: Asbestos was used for fireproofing in pottery made in the Stone Age in the North African Sudan; for insulation in Finnish peasant huts as early as 2,500 B.C.; and in the indestructible wick of lamp before a statue of the goddess Athena in 5th century B.C. Athens.

Pre-and-Post Middle Ages: Plutarch wrote (46-120 A.D.) of an indestructible cloth used for "towels, nets and women's head coverings which cannot be burned by fire; but if any become soiled by use, their owners throw them into the blazing fire and take them out bright and clean." The Emperor Charlemagne (768-814 A. D.) amazed his guests by throwing tablecloths made of this "miracle mineral" into the fire after dinner to be cleaned for the dessert course. Asbestos was used in body armor in the 14th and 15th centuries

Industrial Revolution: The secret of preparing asbestos for weaving was revealed to be boiling the pulverized material in oil. A prosperous industry in textiles, socks, gloves, bags, was set up by Peter the Great around 1720 in Russia, location of the earliest chrysotile or "white" asbestos mines. A sample of chrysotile was given by Benjamin Franklin as a curiosity to the founder of the British Museum in 1725.

Nineteenth Century: Massive chrysotile deposits were discovered in Quebec, Canada, and production soared from fifty tons in 1876 to 21,408 tons by 1900. An asbestos mill was set up by Pope Pius IX in 1830 to produce fireproof paper to protect Papal encyclicals and other documents.

World War I & II: Demand for asbestos to protect the great battleships of the World I British and the American fleets sent Quebec production from a little over 21,408 tons in 1900 to 137,242 tons by the signing of the Armistice. The carriers, destroyers and Liberty ships of World War II needed 345,472 tons in 1940 and 558,181 in 1946 for their hulls and engine rooms.

The Twenties: After a post-war halving of production to 87,457 tons in 1921, the mass-scale manufacture and sale of automobiles sent levels up to 306,055

221

tons by 1929. The demand created a market, for the first time, in crocidolite or "blue" asbestos and amosite or "red" asbestos, mined primarily in South Africa but also in Australia.

First Warnings: Pliny the Elder sounded the first alarm about asbestos' possible health effects in the first century A.D. H. Montagu-Murray reported a case of fibrosis—scarring of the lungs—to the British Departmental Committee on Industrial Disease in 1906. W. O. Cooke, in a 1927 paper for the *British Journal of Medicine*, gave the scarring a name, "asbestosis."

Initial Reaction: European governments in the 1930s laws placed limits on asbestos exposure to workers. The United States passed the Walsh-Healy Act in 1938 providing for the first time federal safety and health protection for workers.

The Emerging Scientific Conflict: In an era when tuberculosis was, with reason, the most feared disease rather than cancer, study of pulmonary processes drew the "best and the brightest" among scientific and the medical students and researchers. The most noted was to become Sir Richard Doll of Oxford University, the epidemiologist who was to firmly establish the cause-and-effect relationship between cigarette smoking and lung cancer.

The Emerging Social Conflict: Labor unions won the right to bargain for hours and wages during the 1930s. But it took a U.S. Supreme Court decision in the late 40s to give the unions the right to negotiate working conditions, including safety and health factors. Asbestos company executives, in the meantime, were exchanging correspondence that suggested deliberate efforts were being made to suppress scientific studies on the adverse health effects of exposure to asbestos at high levels of exposure. And neither they nor the U.S. Navy made any effort to regulate asbestos in shipyards, where dense cloudlike levels were reported...apparently with the approval of President Roosevelt.

Setting the Stage for Conflict: Asbestos production continued to expand after World War II, fueled not only by the demand for the mineral in brake linings, but the hundreds of thousands of new schools, office buildings and shopping malls seeking the fire protection as well as insulating and acoustical properties provided by the mineral. The need for fire protecting materials had been shown by a 1942 fire in the Coconut Grove nightclub in Boston, which claimed almost 500 lives in less than an hour. The demand for insulation grew. Imports of crocidolite and amosite continued to grow. Nevertheless, 95% of the asbestos installed in buildings was chrysolite.

Enter Selikoff: Dr. Irving Selikoff, after being refused funds for research by the asbestos industry, approached unions representing asbestos workers. The result was a study of what appeared to be excessive lung cancer rates in a cohort of 17,800 insulation workers released in the late '60s. The study "blew the asbestos issue out of the quiet circles of scholarly study and into the public consciousness," in the words of a January 18, 1983 *Boston Globe* magazine article. Yet Selikoff himself had carefully qualified his findings with the observation in a 1969 article in the *Journal of the American Medical Science Association*: "Asbestos alone is not the entire explanation. Calculations suggest that asbestos workers who smoke have about 92 times the risk of dying from broncogenic carcinoma (lung cancer) than men who work neither with asbestos nor smoke."

The Environmental Imperative: In 1963, *The New Yorker* magazine published in serial form what was to become the most influential book in American social history of the late 20th century, *Silent Spring* by Rachel Carson. The principal target of the book, the pesticide DDT, was compared to arsenic, "the classic poison of the Borgias." Carson further argued there can be no safe level of exposure to carcinogens, which completely contradicted the universally accepted scientific dose/response principle.

"Sue the Bastards:" The Environmental Defense Fund (EDF) was formed on Long Island, N.Y. in 1968 under the slogan of "Sue the Bastards," and with the express purpose of banning DDT. Among the founding members were Dr. Samuel S. Epstein, later to become the author of *The Politics of Cancer*, called the "Bible of the environmental movement," and Dr. Arthur Upton, who became director of the National Cancer Institute in the Carter Administration. EDF brought cases in communities and states throughout the country seeking a total ban on DDT as a carcinogen.

The Political Imperative: Richard Nixon, taking office after the 1968 Presidential election, the first in modern times in which environmental issues played a dominant role, declared a "War on Cancer." As his first assault on the disease, he created the Environmental Protection Agency (EPA) by executive order. The agency had no overall legislative mandate developed through Congressional hearings—and still does not almost a generation later. The agency operates under a hodgepodge of laws, enacted at different times and with different interpretations—and misinterpretations—of scientific facts since then.

Asbestos Becomes the No. 1 Environmental Enemy: After DDT and two similar pesticides, heptachlor-chlorodane and aldrin-dieldrin, were banned, asbestos once more became a focus of regulatory concern. The permissible exposure level (PEL) belatedly set under the 1938 Walsh-Healey Act in 1969 was reduced from 5 f/cc to 2 by the newly-created (1971) Occupational Safety and Health Administration (OSHA) in 1972-73. EPA soon after prohibited its future use in buildings. But further efforts to ban asbestos entirely had become ensnarled in a one-by-one approach to regulating suspected carcinogens. Consequently, OSHA tried to develop a "generic" regulation policy, which would apply to all substances found to be carcinogenic in animals.

The Lionization of Selikoff: In 1968, *The New Yorker* published a lengthy article about Selikoff written by Paul Brodeur. The article, along with another series by Brodeur in *The New Yorker* in 1973, established Selikoff as the sole authority on asbestos to be trusted by the media—and thereby the public. Further, Brodeur attacked any scientist who dared disagree with Selikoff as being somehow in the pay of industry. Selikoff was portrayed as a champion of public health fighting off the jackals of profit.

The Keller-Kolojeski Shuffle: Two former EPA lawyers who worked closely with Epstein and Upton in the aldrin-dieldrin and heptachlor-cholorodane bans were to be equally important in the development of a highly ambitious attempt to regulate all carcinogens, most of which were still assumed to be produced by industry. Keller left EPA to become OSHA's cancer control expert. Anthony C. Kolojeski established a firm called Clement Associates. Clement Associates was retained by OSHA to provide expert assistance in development of the so-called generic regulation. Selikoff testified on June 1, 1978 in support of the generic policy. He argued that asbestos would cause 40,000 excess deaths from lung cancer, mesothelioma and asbestosis each year. He was contradicted by the National Institute of Occupational Safety and Health (NIOSH), established by Congress to advise OSHA on technical issues. Clement agreed with the proposed OSHA regulation which asserted as much as 20-to-40% of all cancer could be attributed to industrial exposure. NIOSH flatly disagreed, saying only three-to-five percent of cancer could be attributed to industrial exposure and, therefore, the generic policy was not justified nor needed. The result was stalemate.

The Estimates Document: On Sept. 11, 1978, Joseph Califano, then secretary of Health, Education and Welfare (HEW), released a report, which later became known derisively in the scientific community as "The Estimates Document." The document estimated that 2,000,000 deaths could be expected from asbestos over the next three decades. That amounts to 50,000-a-year, 10,000 more than even Selikoff had projected. Asbestos was predicted to be responsible for 17 percent of all cancer deaths. And industrial exposure to nine occupational carcinogens, including arsenic and petroleum distil-lates, to be accountable for 20-to-40 percent of all cancer deaths. Nine names were attached to the estimates document, including those of Upton, Epstein, and Dr. Marvin Schneiderman, later to be become EPA's sole scientific counsel in the 1986-89 proceedings to ban asbestos entirely. But none of the signatories accepted any responsibility for its contents.

The Gathering Storm: The "Estimates Document" was immediately de-nounced in *Science,* the official magazine of the American Association for the Advancement of Science, as "clearly inflated." Similar criticism came from *Lancet,* England's primary scientific publication. Scientists descended upon Washington for a protest meeting. A typical comment at the meeting was that of John Weisburger of the Dana Institute for Cancer Prevention: "[T]his docu-ment would never had seen the light of day if it were ever carefully inspected."

The British Broadside: The calculations in "The Estimates Document" were dismissed as a "confidence trick," in a 1981 paper and book published by Sir Richard Doll in collaboration with Richard Peto, also of Oxford. The estimates, they said, "were so grossly in error that no arguments based even loosely on them should be taken seriously."

The Bureaucratic Counter-Attack: EPA issued a 1982 asbestos-in-schools rule despite the fact that the scientific justification for the rule was scathingly condemned by scientists the agency itself had retained to review the merits of the decision. William Ruckelshaus admitted later the basic principle behind the regulation was to "get the mothers to form a vigilante mob to storm the school committee."

The Scientists Rebel: Dr. Malcolm Ross of the U.S. Geological Survey published in 1984 a review of the scientific literature across the world that established the actual death rate from asbestos was on the order of 520-a-year. And that was at the highest anticipated point, the 1970s when shipyard workers were expected to reach their 60s, and Ross' figures, of course, were only a tiny fraction of Selikoff's projections. Ross' figures were subsequently confirmed by Dr. Robert N. Sawyer, a medical doctor and engineer who directed both the first and largest asbestos removal projects in the country, at Yale University and in the New York City public schools. Sawyer testified before Congressional hearings that "you're causing cancer in abatement workers by this unnecessary removal."

The Foreign Reaction: A Royal Commission in Ontario, Canada concluded in 1984, after months of hearings, that "the health risks to children remain insignificant because the level of exposure in asbestos-containing schools remains extremely low. [T]he program for removing asbestos from all-asbestos-containing schools was not justified by the health risk posed to students." Similar studies and reports in Sweden, France and elsewhere came to the same conclusion.

The Challenger Tragedy: The crash of the Challenger shuttle has been traced directly to a ban on asbestos containing products mandated by the Consumer Product Safety Commission (CPSC). A CPSC ban on asbestos in paints and putties sold the general public prompted by panic over asbestos in hairdryers, led, indirectly, to the discontinuance of the off-the-shelf putty used to seal the O-Rings in the rocket.

A Book Rallies the Scientific Establishment: The publication in 1984 of a book by Edith Efron, entitled *The Apocalyptics: Cancer and the Big Lie: How Environmental Politics Controls What We Know About Cancer* was hailed as "the *Silent Spring* of the counter-revolution." The words were those of Dr. Bruce Ames of the University of California at Berkeley, whose development of a simple, inexpensive test for detecting carcinogens had overthrown the Carson thesis that cancer is a product of the modern, industrial world.

Selikoff Backs Off: From his estimate of 40,000 excess deaths due to asbestos in 1979, Selikoff reduced the figure in half to 20,000 at a Jan. 15, 1980 press conference and then to 8,300 in an interview with *The Boston Globe* a year later. By the fall of 1986, he was saying "if you take averages, there is very little risk" from asbestos in buildings.

Political and Journalistic Perceptions Begin to Change: In March of 1985, *The Detroit News* published a series of articles, "Asbestos Hysteria," written by Michael J. Bennett, author of the present work. The comprehensive series was the first in a major newspaper to challenge the evidence on asbestos, in particular, but also the basic assumptions of the environmental movement: 1. There are only a few carcinogens, usually manmade; and 2. There can be no safe level of exposure to a carcinogen.

Congress Passes a Law: In the Fall of 1986, Congress passed a law sponsored by Rep. James J. Florio (D. - N.J.) called the Asbestos Hazard Emergency Response Act. The law called for all private and public schools in the country to inspect for the presence of asbestos in their buildings by the spring of 1988, develop management plans for its control by the next fall, and have the plans operational by the summer of 1989. The vote in both Houses was unanimous.

Not one dissenting vote was cast. Yet within months, members of both the House and Senate were filing amendments to postpone the application of the deadlines.

Scandals Beset an Unenforceable Law: Within weeks after the passage of AHERA, news organizations were recording scandals associated with a law that was becoming seen as unenforceable as the Volstead Act. Guilty pleas were entered in the case of executives of 42 asbestos abatement firms in New York City indicted for paying off an EPA inspector with $300,000 in bribes. In Denver, a shopping mall was shut down because it was suspected of containing asbestos. The 100 stores in the mall were only allowed to reopen when millions of dollars worth of clothing and fabric, which might contain asbestos drifting down from renovation work, were bagged and dumped. Owners of buildings built between 1935 and 1975 have found the value of their property discounted 15 to 25%. In California, only six out of the 540 contractors doing asbestos control work were found to be registered with the state's Occupational Safety and Health Administration. And, in New Jersey, the president of an asbestos removal firm was gunned down by seven bullets— one in the back of his head.

The Scientific Evidence Becomes Overwhelming: Dr. Morton Corn of Johns Hopkins University, former assistant secretary of Labor, has clearly established asbestos removal compounds health risks by releasing fibers that could otherwise remain safely sealed in place. "[T]he agency has initiated a near-panic response to asbestos in non-occupational settings," Corn said in sworn testimony at an EPA hearing, "and will not acknowledge its overestimate of the magnitude of the risk." A study of a 15-year-old office building in Edmonton, Alberta, Quebec, requested by 2,000 employees has concluded: "The likelihood that a single case of cancer would develop among workers housed in the building as a probably close to zero. . .[and can be] reasonably compared. . .to the probability that someone among the 2,000 would be struck dead by lightning."

Selikoff's Medical Credentials Questioned: The basic medical credentials of Selikoff were questioned in *The Washington Times* on April 21, 1989. Selikoff had informed several biographic directories that he had received his basic medical degree from both the Royal Colleges of Scotland and Melbourne University in Australia in 1941. In fact, while he attended the University of Melbourne, he never received any degree from the institution. Further, he was not awarded his medical degree in Scotland until 1946, five years after embarking on a series of internships and residencies in the United States for which the basic medical degree was a pre-requisite. His only explanation was the comment, "Well, it was the War." Selikoff is a recipient of the Albert Lasker Award, generally regarded as the most prestigious in American science and medicine.

But Congress Wasn't Listening: Rep. Florio introduced more legislation that would apply AHERA to every commercial as well as public building in the country —730,000 in all.

EPA Announces A Total Ban on Asbestos: A total ban on all future uses of asbestos was announced in July of 1989 by EPA Administrator William K. Reilly. He reluctantly conceded the ban would save only 100 to 200 lives over the next 15 years, seven-and-a-half to 15 times less than the agency's original estimate. And 150 of the 200 are brake workers, who are already protected by regulations of the Occupational Safety and Health Administration (OSHA).

Substitutes Are No Substitute: Ross of the U.S. Geological survey, in a paper delivered to the Society of Mining Engineers and the U.S. Stone Association reviewed the scientific literature on substitutes, so-called "vitreous or glassy fibers" in the summer of 1989. He summarized the results of two surveys of 25,000 workers in Europe and 16,000 in the United States as showing "a statistically significant excess of lung cancer (52) percent is seen in these manmade mineral fibers." What is used to replace asbestos may cause more cancer than the asbestos itself.

Smoking Link Stressed: The June 1989 issue of the *New England Journal of Medicine* reiterated the overwhelming influence cigarette smoking plays in inducing asbestos-associated disease. "[I]t remains uncertain whether any type of asbestos acting alone can cause lung cancer in nonsmokers. The article cited five studies shows "no statistically significant cases of lung cancer when exposures are low. It concluded: "In the absence of epidemiological data or estimates of risk that indicates the health risks from environmental exposure are large enough justify high expenditure of public funds, one must question the unprecedented expenses on the order of $100 to $150 billion that could result from asbestos abatement."

Harvard Weighs In: After convening a symposium of the leading asbestos researchers around the world, the Harvard University Energy and Environmental Health Policy Center concluded in August of 1989: "[T]he lifetime risk of premature death due to indoor environmental tobacco smoke (i.e. living with a smoker) or living in a house that has radon is 200 to 400 greater than the projected risks from exposures to asbestos at concentrations" to which school children would be exposed in schools with loose asbestos.

Death from smoking is 20,000 times more likely."

Handwriting on the Wall Time: "We have to get some federal help," Archbishop Theodore E. McCarrick of Newark, N.J. told an assembly of bishops at Seton Hall University in July of 1989. "If we cannot. . .we will be faced with the question of the very survival of parochial schools." And, although public schools, unlike private ones, can pass the expense onto the taxpayer, the bottom line is becoming all too apparent: Money wasted on resolving a phony political crisis, with no scientific foundation in fact, can't be spent on fundamental educational needs. "Asbestos abatement doesn't buy textbooks," as Kate Herber, a spokesman for the National Association of School Boards, observed.

Hypocrisy Self-Exposed: In February, 1990, former U.S. Rep. James J. Florio, the newly elected governor of New Jersey, announced a reduction of $5.9 million in state aid for asbestos removal in schools.

Reality Rears Its Head: About the same time members of Congress were beginning to realize the consequences of AHERA, which it had so blithely passed without a single dissenting vote. For example, one of the two high schools in Caspar, Wyoming was closed down to remove asbestos at a cost of $4 million dollars. A double shift system had to be established at the other school, with one group of students and teachers reporting in the morning and the other in the afternoon. The school now operates from 7 a.m. to 7 p.m. As a consequence, Sen. Malcolm Wallop (R. - Wyoming) has introduced legislation to repeal AHERA, seconded by Sen. Alan Simpson (R. - WY) and Sen. Steve Symms (R. - Idaho), who was instrumental in helping the author of this book establish the role that asbestos—or rather the lack of it—played in the Challenger crash.

An Honest Confession: Sen. Wallop, in introducing his measure to repeal AHERA, said: "The Senate, indeed the Congress has a responsibility to legislate. Too often however, we are a reactive body, passing legislation in haste when we are confronted with a crisis. And here is one of those elegant times. We act without knowledge, without reflection, and it is a poor standard, a terrible example of what we smugly proclaim to be the world's greatest deliberative body. As the students, faculty and parents of Caspar, Wyoming's Kelly-Walsh High School well know, our faults as a legislative body can have costly and disruptive impact on their lives. Fortunately, Congress can undo its mistakes, once it is willing to admit them. That is the purpose of the legislation I am introducing today."

An Admission in Principle:—Grudgingly, On June 12, 1990, less than one year after he announced a total ban on asbestos, EPA Administrator Reilly delivered a speech entitled: "Asbestos, Sound Science, and Public Perceptions: Why we Need a New Approach to Risk." In it he cited as "an excellent example of a clash between real risks and public perception the current controversy over asbestos in the nation's schools and public buildings." He cited, as an inspiration for his talk, a conversation with Sen. Daniel Patrick Moynihan (D-N.Y.) who warned him against allowing, "above all, allowing your agency to become transported by middle-class enthusiasms." Otherwise, the fundamental "law of unintended consequences" would inevitably be invoked. Reilly admitted that asbestos had been a classic case.

An Admission in Fact—Quietly: On Sept. 14, 1990 on a Friday afternoon, a time when the "news hole"—the amount of space and time for news stories on following Saturday—is the tightest for both newspapers and radio-TV, EPA released a guidance document entitled: "Managing Asbestos in Place." None of the mass media was invited, nor any of the specialized publications representing school and commercial building interests, only the newsletters serving the asbestos abatement industry. The document made five key points:

"FACT ONE: Although asbestos is hazardous, the risk of asbestos-related disease depends upon exposure to airborne asbestos fibers. . . . How many fibers a person must breathe to develop disease is uncertain.

"However, at very low exposure levels, the risk may be negligible or zero." That official admission is enough to completely undermine the environmental movement's assertion that there can be no safe level of exposure to asbestos. To rephrase an oft-quoted remark at EPA, "If there is a safe level to asbestos, there can be a safe level to anything."

"FACT TWO: Based on available data, the average airborne asbestos levels in buildings seems to be very low. Accordingly, the health risk to most building occupants seems to be very low.

"FACT THREE: Removal is often *not* the best course of action to reduce asbestos exposure. In fact, an improper removal can create a dangerous situation where none previously existed.

"FACT FOUR: EPA *only* requires asbestos removal in order to prevent significant public exposure to airborne asbestos fibers during building demolition or renovation activities.

"FACT FIVE: EPA *does* recommend a proactive, inplace management program whenever asbestos containing material is discovered.

Only one newspaper, *The Washington Times*, reported the story, under the headline, "EPA Warns Against Asbestos Removal," five days later, September 19, 1990, acting on a tip from the author of this book.

The AMA Speaks and *The New York Times* Takes Note: The American Medical Association on August 6, 1991, advised Americans to "live safely with asbestos" since "it will never be entirely eliminated from the environment." The policy statement added, "The case of asbestos is neither the first, nor is it apt to be the last, of these seeming misconceptions, imbalances and mismatches between scientific fact and the need for action."

"[T]here is a lesson to be learned from the asbestos dilemma," the Association suggested. "Hidsight suggests" if EPA sought scientific consensus in the 80s it "might have better focused and directed its regulatory efforts." *The New York Times* quoted an EPA bureaucrat saying the AMA's "position was quite consistent with EPA's efforts at allaying public misunderstanding and discouraging unneceesary removals."

Why the Farce May Continue: More than 10 years ago, well before EPA announced its original asbestos-in-schools regulation, Dr. William C. Clark, now of Harvard, in an essay titled "Witches, Floods and Wonder Drugs: Historical Perspectives on Risk Management," explained why phenomena such as asbestos hysteria come about—and the rackets that inevitably follow them. There is nothing new about this phenomenon, as Clark observed in comparing the environmental movement's effort to purge the world of carcinogens to the Inquisition's attempt to rid the world of witches: "Witches and witchcraft can be traced back to the very beginnings of history. For centuries, people have found 'witches' a convenient label for their fears of the unknown, an adequate explanation for the inevitable misfortunes which befall one's crops, health and happiness." The Inquisition, in Clark's words, "became the growth industry of the day, offering exciting work, rapid advancement and wide recognition to its professional and technical workers." As Clark wrote, "By the dawn of the Enlightment, witches had been virtually eliminated from Europe and North America." But it was public revulsion at the Inquisition itself that caused it. An Inquisitor named Lanes Caldera y Fries in 1610 showed that most of the accusations in a century of witchcraft trials were false. Torture and the presumption of guilt had created witches where none existed. And there was not a single case of witchcraft to show for all the preaching, hunting and burning that had been carried out in the name of the church. Today asbestos is the new witchcraft. Environmentalists are the new Inquisitors. Journalists are the town criers who help the new Inquisitors preach, hunt and burn the witches of asbestos. At the time AHERA was passed by Congress, "everyone knew" asbestos was the most dangerous carcinogen around. That was everyone except the scientists—because, tragically for both journalism and the country, reporters all too often are not fairminded skeptics but governmental propagandists.

Index

A

Abelson, Philip, 171.
AFL-CIO, 157.
Agency for International Development (AID), 51.
Agent Orange, 129, 131, 197-201, 203-205.
Albert Lasker Award, 93, 228.
"America the Beautiful" (song), 6.
American Association for the Advancement of Science, 34, 121, 213, 225.
American Enterprise Institute, 13, 43.
American Men and Women of Science, 94.

Ames, Dr. Bruce, 61-62, 115-116, 124-125, 213, 215, 226.
Amosite, 22-23, 90, 92, 102, 105, 111, 113-114, 146, 151, 176, 222.
Amphiboles, 14, 17-18, 23, 90.
Anderson, Adelaide, 94.
Anderson, Arthur, 89.
Anderson, Robert C., 87.
The Apocalyptics: Cancer and the Big Lie, 99, 115, 158, 215, 226.
Aquinas, St. Thomas, 215.
Arco, 28.
Arabic, 215.
Aramid, fiber, 85.
Aristotle, 6.
Arnett, Peter, 9.

Arsenic, 24, 62, 162, 223, 225.
Asbestos ,1, 8-10, 13-29, 31-38, 41, 43-62, 65-69, 75-79, 85-87, 89-99, 101-114, 115-116, 119, 122, 126, 131-135, 137-153, 157, 163-164, 166, 168, 170-171, 175-184, 187-195, 197-201, 205, 209, 211-212, 214, 218, 221-231.
Asbestos abatement, 1, 19-21, 27-28, 33, 44, 49, 56, 135, 151, 227-230.
Asbestos and Disease, 90, 92.
Asbestos-containing materials, 17-20, 142-143, 145, 148, 178, 206.
Asbestos Hazard Emergency Response Act (AHERA), 10, 17, 19, 26, 45, 55, 182, 205, 211, 218, 227-229, 231.
Asbestosis, 23, 94, 98-99, 103, 110, 114, 190, 222, 224.
Asbestos issues, 29.
Asbestos-related malignancy, 21.
Asbestos workers, 19, 21, 53, 58, 101, 104, 113, 147, 164, 170-171, 181, 223.
Associated Press, 9, 179.
Auchter, Thorne, 197-198.
Audubon Society, 34.
Axelrod, David, 122.

B

Baier, Edward, 132, 165.

233